SMALL BUSINESS ADMINISTRATION: OPERATIONS AND PROGRAMS

SMALL BUSINESS ADMINISTRATION: OPERATIONS AND PROGRAMS

U.S. GOVERNMENT ACCOUNTABILITY OFFICE

Nova Science Publishers, Inc.
New York

For permission to use material from this book please contact us:

Telephone 631-231-7269; Fax 631-231-8175

Web Site: http://www.novapublishers.com

LIBRARY OF CONGRESS CATALOGING-IN-PUBLICATION DATA

United States. Government Accountability Office.

Small Business Administration : operations & programs / GAO.

p. cm.

ISBN 978-1-60692-043-5 (softcover)

1. United States. Small Business Administration--Management. 2. Small business--Government policy--United States. 3. Small business--United States--Finance. 4. Federal aid to small business--United States. I. Title.

HD2346.U5U53 2008

354.2'7990973--dc22 2008031083

Published by Nova Science Publishers, Inc. ✢ New York

CONTENTS

PREFACE

The Small Business Administration's (SBA) 7(a) program, initially established in 1953, provides loan guarantees to small businesses that cannot obtain credit in the conventional lending market. In fiscal year 2006, the program assisted more than 80,000 businesses with loan guarantees of nearly $14 billion. This book examines (1) the program's purpose, based on its legislative history, and performance measures; (2) evidence of constraints, if any, affecting small businesses' access to credit; (3) the types of small businesses served by 7(a) and conventional loans; and (4) differences in SBA's estimates and reestimates of the program's credit subsidy costs. GAO analyzed agency documents, studies on the small business lending market, and data on the characteristics of small business borrowers and loans.

Chapter 1

RESULTS IN BRIEF[*]

The 7(a) program's design and performance measures in part reflect the program's legislative history, but the performance measures provide limited information about the impact of the loans on the small businesses receiving them. The underlying statutes and legislative history of the 7(a) program help establish the federal government's role in assisting and protecting the interests of small businesses, especially those with minority ownership. The program's performance measures focus on loan guarantees that are provided to small business owners identified in the program's authorizing statutes and legislative history. These firms include start-ups, existing small businesses, and businesses whose owners face "special competitive opportunity gaps," such as minority- or female-owned businesses. However, all of the 7(a) program's performance indicators are primarily output measures—for instance, they report on the number of loans approved and funded. As a result, no information is available on how well firms do after receiving a 7(a) loan (outcomes). The current measures do not indicate how well the agency is meeting its strategic goal of helping small businesses within these groups succeed. The agency is currently undertaking efforts to develop additional outcome-based performance measures for the 7(a) program, but agency officials said that it was not clear when any outcome-based measures might be introduced or what they might measure.

Limited evidence from economic studies suggests that some small businesses may face constraints in accessing credit because of imperfections, such as credit rationing, in the conventional lending market. Some studies

---

[*] Excerpted from GAO Report GAO-07-769, dated July 2007.

showed, for example, that lenders might lack the information needed to distinguish between creditworthy and noncreditworthy borrowers and thus could "ration" credit by not providing loans to all creditworthy borrowers. Several studies we reviewed generally concluded that credit rationing was more likely to affect small businesses because lenders could face challenges in obtaining enough information on these businesses to assess their risk. The literature we reviewed on credit rationing relied on data from the early 1970s through the early 1990s, however, and did not account for recent trends in the small business lending market. Among these trends is the increased use of credit scoring, which provides lenders with additional information on borrowers and may have had a significant impact on the extent of credit rationing in the current conventional lending market. In addition to credit rationing, some lenders may deny credit to firms owned by specific segments of society. Though studies we reviewed noted some disparities among races and genders in the conventional lending market, the studies did not offer conclusive evidence on the reasons for those differences.

7(a) loans went to certain segments of the small business lending market in higher proportions than conventional loans. For example, 28 percent of 7(a) loans compared with an estimated 9 percent of conventional loans went to minority-owned small businesses from 2001 through 2004. In addition, 25 percent of 7(a) loans went to small business start-ups, while the overall lending market served almost exclusively established firms (about 95 percent). A more similar percentage of 7(a) and conventional loans went to other segments of the small business lending market, such as businesses owned by women or located in distressed neighborhoods. Finally, the characteristics of 7(a) and conventional loans differed in several ways. For example, 7(a) loans typically were larger and more likely to have variable rates, longer maturities, and higher interest rates than conventional loans to small businesses.

SBA's most recent reestimates of the credit subsidy costs for 7(a) loans made during fiscal years 1992 through 2004 indicate that, in general, the long-term costs of these loans would be lower than initially estimated. The 7(a) program has been estimated to be a "zero credit subsidy" program since fiscal year 2005. The most recent reestimates, including those made since 2005, may change because of the inherent uncertainties of forecasting subsidy costs and the influence of economic conditions, such as interest rates on several factors, including loan defaults (which exert the most influence over projected costs) and prepayment rates. Unemployment is another factor related to the condition of the national economy that could affect the credit subsidy cost—for instance, if unemployment rises above projected levels, loan defaults are likely to increase. Beginning in 2003, the agency has moved from primarily using historical averages of loan performance data to an econometric model that incorporates historical data and other economic assumptions to project credit subsidy costs.

This report makes a recommendation to the SBA Administrator to complete and expand SBA's current work on evaluating the program's performance measures. In addition, we recommend that SBA use the loan performance information it already collects, including but not limited to defaults, prepayment rates, and the number of loans in good standing, to better report how small businesses fare after they participate in the 7(a) program.

We provided a draft of this report to SBA for review and comment. In written comments, SBA agreed with our recommendation (see app. IV). However, SBA disagreed with a comparison in the section of our report discussing credit scores of borrowers with 7(a) and conventional loans. Specifically, we reported limited differences in the credit scores of small businesses with 7(a) and conventional loans. Although stating in its letter that "the numbers have not been worked out," SBA concluded that the impact on loan defaults from the higher share of 7(a) loans in the riskier credit score categories would not be insignificant. Our analyses of credit scores and other borrower and loan characteristics was not intended to quantify the impact of differences in these characteristics on 7(a) defaults. We continue to believe that our analysis of credit scores provides a reasonable basis for comparing the scores of business in different credit score categories. Further analyses of these types are consistent with our recommendation that SBA expand its abilities to assess the overall effectiveness of the 7(a) program. In addition, SBA provided technical comments, which we incorporated into the report as appropriate.

Chapter 2

BACKGROUND

Initially established in 1953, the 7(a) program guarantees loans made by commercial lenders—mostly banks—to small businesses for working capital and other general business purposes.[1] The guarantee assures the lender that if a borrower defaults on a loan, the lender will receive an agreed-upon portion (generally between 50 percent and 85 percent) of the outstanding balance. Because the guarantee covers a portion of the outstanding amount, both the lender and SBA share some of the risk associated with a potential default. SBA is not liable for the guarantee should the lender not comply materially with the program's regulations—for instance, by not paying the guarantee fee to SBA in a timely manner. As figure 1 shows, SBA's share of loans guaranteed by the 7(a) program was an estimated 4.1 percent of all outstanding small business loan dollars for loans under $1 million ($24.7 billion out of $600.8 billion). This share accounts for about 1.3 percent of the number of outstanding small business loans of under $1 million in 2005 (about 264,000 out of 21 million loans).[2] SBA's shares of outstanding small business loans under $1 million for the years 2003 and 2004 were similar.[3]

SBA relies on lenders to process and service 7(a) loans and to ensure that borrowers meet the program's eligibility requirements.[4] To be eligible for the 7(a) loan program, a business must be an operating for-profit small firm (according to SBA's size standards) located in the United States. To determine whether a business qualifies as small for the purposes of the 7(a) program, SBA uses size standards that it has established by industry.[5] These standards set the maximum average number of employees or annual receipts that a small business may have. While SBA gives special consideration to certain groups of business owners, the program does not set aside loans for or require that a certain number of loans be made to targeted groups. Nevertheless, SBA has performance measures that track how many loans go to new small businesses and that include information on various types of businesses, such as minority-, women-, and veteran-owned firms.

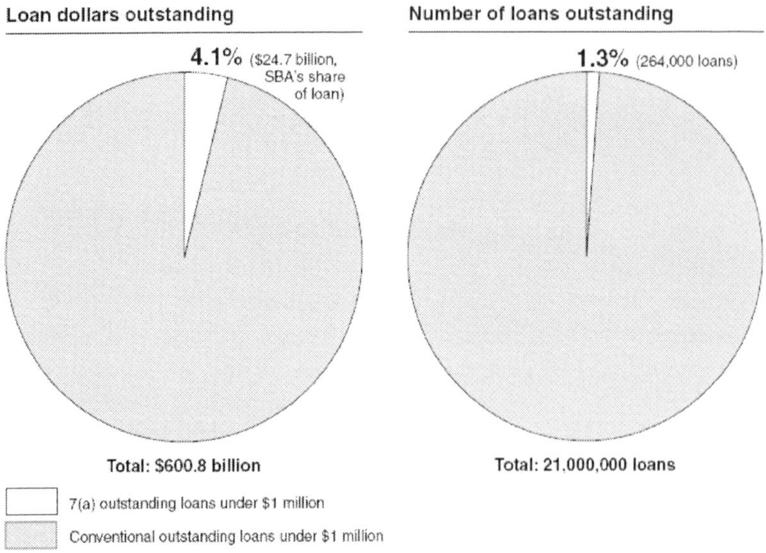

Source: GAO analysis of SBA outstanding 7(a) loan data and Office of Advocacy special tabulations of call reports (Consolidated Reports of Condition and Income for U.S. Banks).

Figure 1. Loan Volume for 7(a) and Conventional Small Business Loans, 2005.

In addition to making sure that borrowers meet the size requirements, lenders must certify that small businesses meet the "credit elsewhere" requirement. SBA does not extend credit to businesses if the financial strength of the individual owners or the firm itself is sufficient to provide or obtain all or part of the financing or if the business can access conventional credit. To certify borrowers as having met the credit elsewhere requirement, lenders must first determine that the firm's owners are unable to provide the desired funds from their personal resources. Second, the credit elsewhere test requires that lenders determine that the desired credit, for similar purposes and period of time, is unavailable to the firm on reasonable terms and conditions from nonfederal sources without SBA assistance, taking into consideration prevailing rates and terms in the community or locale where the firm conducts business. Nonfederal sources may include any lending institutions or a borrower's personal resources.

According to SBA's fiscal year 2003-2008 Strategic Plan, the agency's mission is to maintain and strengthen the nation's economy by enabling the establishment and viability of small businesses and by assisting in the economic recovery of communities after disasters. SBA describes the 7(a) program as contributing to an agencywide goal to "increase small business success by bridging competitive opportunity gaps facing entrepreneurs." As reported annually in SBA's Performance and Accountability Reports (PAR), the 7(a) program contributes to this strategic goal by fulfilling each of the following three long-term, agencywide objectives: (1) increasing the positive impact of SBA assistance on the number and success of small business start-ups, (2) maximizing the sustainability and growth of existing small businesses that receive SBA assistance, and (3) significantly increasing successful small business ownership within segments of society facing special competitive opportunity gaps. Groups facing these special competitive opportunity gaps include those that SBA considers to own and control little productive capital and to have limited opportunities for small business ownership (such as African Americans, American Indians, Alaska Natives, Hispanics, Asians, and women) and those that are in certain rural or low-income areas. The 7(a) program has nine performance measures. For each of its three long-term objectives, SBA collects and reports on (1) the number of loans approved, (2) the number of loans funded (i.e., money that was disbursed), and (3) the number of firms assisted.

To report on its performance measures, SBA collects data from lenders. Loan-level data for the 7(a) program are housed in the Loan Accounting System. This system contains data describing the loan, such as the percentage of the loan guaranteed by SBA, the number of months to maturity, and the interest rate (fixed or variable). The data also include information on the small firm, such as the ethnicity and gender of the principal owner, the number of employees, and the firm's status as "new" (i.e., less than 2 years old). Furthermore, the system contains data on the loan's status—for example, whether the loan has been purchased by SBA (i.e., is in default), has been prepaid, or is in good standing.

According to provisions in FCRA, at the time a guaranteed loan is made, the credit subsidy cost is financed with the program's annual appropriations. Also under FCRA, SBA makes annual revisions (reestimates) of credit subsidy costs for each cohort of loans made during a given fiscal year using new information about loan performance, revised expectations for future economic conditions and loan performance, and improvements in cash flow projection methods. These reestimates represent additional costs or savings to the government and are recorded in the budget. FCRA provides permanent indefinite budget authority for any reestimated increases of credit subsidy costs (upward reestimates) that occur after the year in which a loan is disbursed. Reestimated reductions of subsidy costs (downward reestimates) are credited to the Treasury and are unavailable to the agency. In addition, FCRA does not count administrative expenses against the appropriation for credit subsidy costs. Instead, administrative expenses are subject to separate appropriations and are recorded each year as they are paid, rather than as loans are originated.

Chapter 3

THOUGH INCORPORATING POLICY OBJECTIVES FROM THE 7(A) PROGRAM'S LEGISLATIVE HISTORY, 7(A)'S PERFORMANCE MEASURES DO NOT GAUGE THE PROGRAM'S IMPACT ON PARTICIPATING FIRMS

The performance measures for the 7(a) program incorporate the various policy objectives described in the program's underlying statutes and legislative history but do not assess the impact of the loan guarantees on small businesses receiving loans. We compared criteria for the characteristics of effective performance measures and found that the 7(a) performance measures incorporated several of these attributes. For example, the performance measures track the main activity of the 7(a) program by identifying the number of loans that are approved for small firms that have been unable to obtain credit in the conventional lending market. However, the performance measures do not show whether the program is meeting the agency's goal of improving the success of small firms that participate in the program. None of the 7(a) performance measures provide information on how well firms do after they have received a loan. SBA has been undertaking efforts to develop additional performance measures to describe the program's impact on participating firms. But the agency has yet to define specific outcome-based performance measures and does not have a time line for implementing such measures.

The 7(a) Program's Legislative History Emphasizes the Program's Role in Meeting Credit Needs of Certain Small Businesses

The 7(a) program's underlying statutes and legislative history have helped establish the federal government's role in assisting and protecting the interests of small business, taking into account the importance of these businesses to the overall functioning of the national economy. The legislative basis for the 7(a) program recognizes that the conventional lending market is the principal source of financing for small businesses and that the loan assistance that SBA provides is intended to supplement rather than compete with that market. However, as the legislative history suggests, conventional lending may not be a feasible financing option for some small businesses under certain circumstances. For example, conventional lenders may be unwilling to make loans when the risk of a small business is difficult to assess—for instance, when they believe that the small business has insufficient assets or specialized inventory and equipment or lacks a credit history, as in the case of a start-up. In addition, the loan terms offered to a small business in the conventional lending market may not be practical—for example, a small business may need loans with longer-term maturities than conventional lenders may be willing to provide.

The design of the 7(a) program is consistent with the program's underlying statutes and legislative history in that SBA collaborates with the conventional market in identifying and supplying credit to small businesses in need of assistance. Specifically, the 7(a) program has three design features that help it address concerns identified in its legislative history. First, the loan guarantee, which plays the same role as collateral, limits the lender's risk in extending credit to a small firm that may not have met the lender's own requirements for a conventional loan. According to SBA officials, a lender's willingness to underwrite the loan only with the guarantee confirms that the 7(a) program fills a credit gap. Second, the "credit elsewhere" requirement is intended to provide some assurance that guaranteed loans are offered only to firms that are unable to access credit on reasonable terms and conditions in the conventional lending market. Lenders follow SBA policies and procedures in determining whether a small business fulfills this key 7(a) program requirement. SBA officials explained that the agency is currently reviewing how lenders apply the credit elsewhere requirement, though the results of this review are not yet complete. Third, an active secondary market for the guaranteed portion of a 7(a) loan allows lenders to sell the guaranteed portion of the loan to investors, providing additional liquidity that lenders can use for additional loans.

Numerous amendments to the Small Business Act and to the 7(a) program have laid the groundwork for broadening small business ownership among certain groups, including veterans, handicapped individuals, women, African Americans, Hispanics, Native Americans, and Asians. The 7(a) program also includes provisions for extending financial assistance to small businesses that are located in urban or rural areas with high proportions of unemployed or low-income individuals or that are owned by low-income individuals. The program's legislative history highlights its role in helping small businesses, among other things, get started, allowing existing firms to expand, and enabling small businesses to develop foreign markets for their products and services.

The 7(a) Program's Performance Measures Are Related to the Program's Core Activity, but Do Not Provide Information on Its Impact on Participating Firms

We stated in earlier work that a clear relationship should exist between an agency's long-term strategic goals and its program's performance measures.[6] Outcome-based goals or measures showing a program's impact on those it serves should be included in an agency's performance plan whenever possible. Most plans typically supplement outcome goals with output goals showing the number and type of services provided because the program may not meet an outcome goal in the year covered by the plan. In some cases, a goal may be too difficult to measure. In previous work, we have also identified specific attributes of successful performance measures.[7] For example, each performance measure should have a measurable target and explicit methodology showing how that target was determined. Without a measurable target, an organization may be unable to determine whether it is meeting its goals. Table 1 provides a detailed description of these key attributes and discusses the potentially adverse consequences of not incorporating them into performance measures.

Table 1. Attributes of Successful Performance Measures

Attribute	Definitions	Potentially adverse consequences of not meeting attribute
Core program activity	Measure covers the activities that an entity is expected to perform in support of the program's intent.	Managers and stakeholders may not have enough information in core program areas.
Measurable target	Measure has a numerical goal.	It may be impossible to determine whether a program's performance is meeting expectations.
Reliability	Measure produces the same result under similar conditions.	Reported performance data are inconsistent and uncertainty exists about them.
Clarity	Measure is clearly stated and the name and definition are consistent with the methodology used to calculate it.	Data could be misleading to users and not capture what is intended to be measured.
Objectivity	Measure is reasonably free from significant bias or manipulation.	Performance assessments may be systematically over-or understated.
Linkage	Measure is aligned with division and agencywide goals and mission.	Behaviors and incentives created by measures do not support achieving division or agencywide goals or mission.

Source: GAO-03-143.

We reviewed SBA's performance measures for the 7(a) loan program and found that the measures generally exhibited all of the traits described above, except for the measurable target and linkage attribute. According to SBA's fiscal year 2006 PAR, the nine performance measures were:

1 number of new loans approved to start-up small businesses,
2 number of new loans funded to start-up small businesses,
3 number of start-up small businesses assisted,
4 number of new loans approved to existing small businesses,
5 number of new loans funded to existing small businesses,
6 number of existing small businesses assisted,
7 number of new loans approved to small businesses facing special competitive opportunity gaps,
8 number of new loans funded to small businesses facing special competitive opportunity gaps, and

9 number of small businesses facing special competitive opportunity gaps assisted.

All nine performance measures we reviewed provided information that related to the 7(a) loan program's core activity, which is to provide loan guarantees to small businesses. In particular, the indicators all provided the number of loans approved, loans funded, and firms assisted by subgroups of small businesses the 7(a) program is intended to assist. As stated earlier, the program's legislative history indicates that SBA's specific lending objectives include stimulating small business in distressed areas, promoting small businesses' contribution to economic growth, and promoting minority enterprise opportunity. Consequently, SBA has developed performance measures that specifically track how many guaranteed loans go to those small business owners that the agency refers to collectively as facing special competitive opportunity gaps. Similarly, SBA separately tracks loan data regarding start-up small businesses, another group that the 7(a) program's legislative history specifically cites as having challenges in obtaining credit within the conventional lending market.

As table 2 shows, in 2004 and 2005 SBA generally met or exceeded its goals for the number of loans approved for start-ups, existing small businesses, and businesses facing special competitive opportunity gaps. In 2006, SBA did not meet any of its targets for these measures. However, while the 7(a) program did not meet its targets, it approved slightly more than 90 percent of the loans that it had set as its goal. SBA also did not always meet its target for the number of firms assisted. In years when SBA did not meet these targets, the 7(a) program again met almost 90 percent of its goal for firms assisted. Though it is not clear why SBA did not meet these targets, SBA's fiscal year 2006 PAR suggests that there may have been less demand for 7(a) loans. In addition, SBA officials explained that the agency did not make loans to small businesses directly and therefore had less control over the number of loans made. Instead, the agency relies primarily on marketing and community outreach to inform both lenders and prospective borrowers about the 7(a) program. Furthermore, SBA officials explained that the 7(a) program staff leverages other SBA offices, such as those that offer technical assistance to small businesses, to further raise the awareness among the general public and potential lenders about the 7(a) program.

Table 2. 7(a) Performance Measure Targets and Results, 2004-2006

	Fiscal year					
	2004		2005		2006	
Performance measures	Target	Result	Target	Result	Target	Result
Number of loans approved						
Start-up small business	18,000	18,134	22,671	29,587	33,024	32,983
Existing small business	72,000	62,999	65,305	66,313	73,536	64,307
Small business facing special competitive opportunity gap	44,617	60,787	68,621	74,307	76,690	71,326
Number of firms assisted						
Start-up small business	18,000	15,351	22,671	25,086	28,224	27,368
Existing small business	72,000	53,544	65,305	57,296	62,144	52,935
Small business facing special competitive opportunity gap	44,617	52,075	68,621	64,390	64,377	60,691

Source: GAO analysis of SBA's fiscal years 2006 and 2007 Budget Request and Performance Plan and fiscal year 2006 PAR.

By having quantifiable goals, all of the performance measures partly met our criterion for having a measurable target attribute. SBA annually reports performance measure data, publishing goals in the agency's annual Budget Request and Performance Plan for the upcoming fiscal year and results for the preceding fiscal year in its PAR.

Though having measurable targets is a positive attribute, the PAR does not contain information about how SBA set its goals. According to SBA officials, the actual targets set for all of the measures related to the 7(a) program are based on historical data. SBA officials explained that the number of loans approved is calculated by dividing the amount appropriated for loan guarantees in a given fiscal year by the previous fiscal year's average loan amount, producing a target for the number of loans approved. SBA also measures the number of loans funded and firms assisted, both of which closely track the number of loans approved. According to SBA officials, both of these measures are always slightly lower than the number of loans approved because not all approved loans are funded and the number of firms assisted does not include multiple loans to the same firm in a given fiscal year.

In addition, the 7(a) program's performance measures are generally reliable, clearly defined, and objective. Our assessment of SBA's databases that contain information on the agency's performance measures concluded that these data were sufficiently reliable for the purposes of evaluating key loan characteristics. Additionally, most of the measures are clearly described in the SBA documents that addressed the 7(a) program's performance measures, since each performance measure's name is also its definition. Finally, the performance measures are objective and generally free from any biases, in part because they simply report the overall annual volume (i.e., outputs) of guaranteed lending business.

Since all of the 7(a) program's performance measures are primarily output measures—that is, they report on the number of loans approved and funded and firms assisted—SBA does not collect any information that discusses how well firms are doing after receiving a 7(a) loan (outcomes). Further, none of the measures link directly to SBA's long-term objectives. As a result, the performance measures do not fully support SBA's strategic goal to "increase small business success by bridging competitive opportunity gaps facing entrepreneurs." We noted in 1999 that SBA relies on output measures, such as an increase in the number of loans, but does not show how these measures are related to increasing opportunities for small businesses to be successful—SBA's main goal.[8] SBA's Inspector General also concluded in a 2000 report that most 7(a) performance measures were output based and did not provide information showing the extent to which the program was accomplishing its mission under the Small Business Act.[9] SBA management concurred with the Inspector General's conclusion and recommendations, including that the agency develop performance measures to gauge outcomes and goals for meeting the requirements set forth in the Government Performance and Results Act of 1993 (GPRA).

SBA IS WORKING TO GAUGE THE 7(A) PROGRAM'S IMPACT ON PARTICIPATING FIRMS

SBA officials have recognized the importance of developing performance measures that better assess the 7(a) program's impact on the small firms that receive the guaranteed loans. SBA is expecting a final report in the summer of 2007 from the Urban Institute, which has been contracted to undertake several evaluative studies of several programs, including 7(a), that provide financial assistance to small businesses. Components of this work include assessing potential duplication of SBA's main financial assistance programs by state or local programs, establishing a baseline measure of SBA customer satisfaction, and interviewing participating lenders about their underwriting practices. One component of the study that will not be undertaken is an analysis to determine how outcomes for firms assisted through financial assistance programs, such as 7(a), would differ in the absence of SBA assistance. The impact study, as designed by the Urban Institute, required the use of credit scores for firms that did not receive SBA assistance.[10] Though costs associated with this component of the study initially prohibited SBA from undertaking it, SBA officials explained that they were advised that they are legally prohibited from obtaining credit score data from firms with which they have no relationship.

SBA officials explained that no formal decision had yet been made about how the agency might alter or enhance the current set of performance measures to provide more outcome-based information related to the 7(a) program, for several reasons. These included the agency's reevaluation of its current strategic plan in response to GPRA's requirement that agencies reassess their strategic plans every 3 years, a relatively new administrator who may make changes to the agency's performance measures and goals, and the cost and legal constraints associated with the Urban Institute study.[11] However, SBA already collects information showing how firms are faring after they obtain a guaranteed loan. In particular, SBA regularly collects information on how well participating firms are meeting their loan obligations. This information generally includes, among other things, the number of firms that have defaulted on or prepaid their loans—data that can serve as reasonable proxies for determining a firm's financial status. Though this information provides some indication of how successful firms are after receiving a 7(a) guaranteed loan, the agency primarily uses the data only to estimate some of the costs associated with the program and for internal reporting purposes, such as monitoring participating lenders and analyzing its current loan portfolio. Expanding uses of this information as part of its performance measures could provide SBA and others helpful information for describing the financial status of firms that have been assisted by the 7(a) program.

Chapter 4

LIMITED EVIDENCE SUGGESTS THAT CERTAIN MARKET IMPERFECTIONS MAY RESTRICT ACCESS TO CREDIT FOR SOME SMALL BUSINESSES

Limited evidence from economic studies suggests that some small businesses may face constraints in accessing credit because of imperfections, such as credit rationing, in the conventional lending market. But this evidence is based on data that end with the early 1990s and do not account for developments that have occurred in the small business lending market since then. We focused on evidence of credit rationing reported in academic studies published in peer-reviewed journals.[12] With some exceptions, the studies we reviewed generally concluded that credit rationing was more likely to exist when there was a lack of information about the borrower—for example, with small businesses—and that the effect of this type of credit constraint on the national economy was not likely to be significant. However, the research on credit rationing was limited by at least two factors. First, researchers do not all use a similar definition for credit rationing. Second, as we have noted the studies we reviewed did not consider recent developments in the small business lending market, such as the increasing use of credit scores, that may reduce credit rationing. Finally, though researchers have noted disparities in lending options among different races and genders, inconclusive evidence exists as to whether discrimination explains these differences.

Studies We Reviewed Provide Limited Evidence of Credit Rationing

We found limited information that credit constraints, such as credit rationing, could have some effect on small businesses. Credit rationing, or denying loans to creditworthy individuals and firms, generally stems from lenders' uncertainty or lack of information regarding a borrower's ability to repay debt. Economic reasoning suggests that there exists an interest rate (i.e., the price of the loan) beyond which banks will not lend, even though there may be creditworthy borrowers willing to accept a higher interest rate.[13] Because the market interest rate will not climb high enough to convince lenders to grant credit to these borrowers, these applicants will be unable to access credit and will also be left out of the lending market.[14] Of the studies we identified that empirically looked for evidence of credit rationing within the conventional U.S. lending market, almost all provided some evidence consistent with credit rationing.[15] For example, one study found evidence of credit rationing across all sizes of firms.[16] However, another study suggested that the effect of credit rationing on small firms was likely small, and another study suggested that the impact on the national economy was not likely to be significant.[17] Specifically, one of these two studies, which used data on small businesses, concluded that though crediting rationing was associated with firm size, it was economically unimportant to the small businesses within their dataset.[18] Only one study that we reviewed found no evidence of credit rationing, though it could not rule out the existence of this market imperfection.[19]

In some studies we reviewed, we also found that researchers used different definitions of credit rationing and that a broader definition was more likely to yield evidence of the existence of credit rationing than a narrower definition. For example, one study defined a firm as being credit rationed if the firm was either denied a loan or discouraged from applying for credit.[20] However, another study pointed out that firms could be denied credit for reasons other than credit rationing, such as not being creditworthy.[21] Because the underlying reason for having been denied credit can be difficult to determine, true credit rationing is difficult to measure.

Other studies of small business lending that we reviewed found evidence for credit rationing by testing whether the circumstances of denial were consistent with a "credit rationing" explanation, such as a lack of information. For example, two studies concluded that having a preexisting relationship with the lender had a positive effect on the borrower's chance of obtaining a loan.[22] The empirical evidence from another study suggested that lenders use information accumulated over the duration of a financial relationship with a borrower to define loan terms. This study's results suggested that firms with longer relationships received more favorable terms—for instance, they were less likely to have to provide collateral. Because having a relationship with a borrower would lead to the lender's having more information, the positive effect of a preexisting relationship is consistent with the theory behind credit rationing.[23]

Aside from credit rationing, lenders could potentially deny creditworthy firms a loan because of the race or gender of the owner. This practice would also constitute a market imperfection because lenders would be denying credit for reasons other than interest rate or another risk associated with the borrower. A 2003 survey of small businesses conducted by the Federal Reserve examined differences in credit use among racial groups and between genders. The survey found that differences did not exist across all comparison groups.[24] For example, the survey found that 48 percent of small businesses owned by African Americans and women and 52 percent of those owned by Asians had some form of credit, while 61 percent of white-owned or Hispanic-owned businesses had some form of credit.[25]

Studies have attempted to determine whether such disparities are due to discrimination, but the evidence from the studies we reviewed was inconclusive. For example, one study found evidence that discrimination existed against Hispanics and Asians, but not against African Americans and women.[26] A different study that was able to control for the effects of a variety of variables, such as whether the borrower had experienced bankruptcy and the borrower's credit score, found some evidence of discrimination against African Americans and women, but not against other minorities.[27] Finally, a third study found significant evidence that only firms owned by African Americans faced obstacles in obtaining credit and were charged higher interest rates, while the study did not find significant evidence that other minority- and women-owned firms face discrimination.[28]

THE LITERATURE DOES NOT ADDRESS RECENT TRENDS IN THE SMALL BUSINESS LENDING MARKET

The studies we reviewed regarding credit rationing used data from the early 1970s through the early 1990s and thus did not account for several recent trends that may have impacted the extent of credit rationing within the small business lending market. According to a Federal Reserve report on the availability of credit for small businesses, lenders are increasingly using credit scores in loan decisions involving small businesses. Credit scores provide additional information about borrowers and may reduce the cost to lenders of evaluating the risk potential borrowers present. As a result, credit scores may decrease the extent to which credit rationing occurs. Further, our economic literature review identified one study suggesting that the recent changes in bankruptcy laws may also impact the small business lending market because loans to small businesses are often secured by personal credit. Specifically, the change in bankruptcy laws that occurred in October 2005 may have made it more difficult for some individuals to declare bankruptcy and thus decreased the risk to lenders, making lenders more willing to extend credit. In addition, because it has become harder to declare bankruptcy, potential borrowers may be less likely to apply for a loan. These trends may also lead to less credit rationing in the conventional lending market. Finally, considerable consolidation has taken place in the banking industry and may have led to a decrease in the number of small banks. Historically, smaller banks have been more involved with small business lending because of the relationships between small local banks and local firms. As noted previously, relationships with lenders can limit credit rationing. With the potential decline in the number of small banks, these relationships may diminish, possibly leading to more credit rationing.

Chapter 5

A HIGHER PERCENTAGE OF 7(A) LOANS WENT TO CERTAIN SEGMENTS OF THE SMALL BUSINESS LENDING MARKET, BUT CONVENTIONAL LOANS WERE WIDELY AVAILABLE

7(a) loans went to certain segments of the small business lending market in higher proportions than conventional loans. From 2001 to 2004, a higher percentage of 7(a) loans went to minority-owned and start-up businesses compared with conventional loans. However, more similar percentages of loans with and without SBA guarantees went to small businesses owned by women and those located in economically distressed neighborhoods. The characteristics of 7(a) and market loans differed in several key respects. For example, loans guaranteed by the 7(a) program were more likely to be larger and have variable interest rates, longer maturities, and higher interest rates.

HIGHER PROPORTION OF 7(A) LOANS WENT TO MINORITY-OWNED AND START-UP SMALL BUSINESSES

From 2001 to 2004, minority-owned small businesses received a larger share of 7(a) than conventional loans. More than a quarter of 7(a) loans went to small businesses with minority ownership, compared with an estimated 9 percent of conventional loans (figure 2). However, in absolute numbers many more conventional loans went to the segments of the small business lending market we could measure, including minority-owned small businesses, than loans with 7(a) guarantees. For example, if we apply the percentage of 7(a) loans going to minority-owned firms (28 percent) from 2001 through 2004 to the number of outstanding 7(a) loans under $1 million in 2004 (223,939), an estimated 62,000 of these outstanding 7(a) loans went to minority-owned small firms. In comparison, if we apply the percentage of conventional loans going to minority-owned firms over the same period (9 percent) to the number of outstanding loans under $1 million in 2004 (17.13 million), we estimate that there were more than 1.6 million outstanding loans to minority-owned small businesses in June 2004.

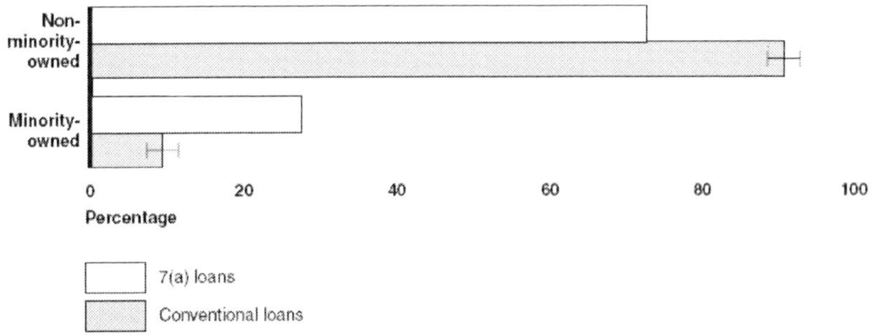

Source: GAO analysis of SBA and Federal Reserve Board of Governors' data.

Note: The brackets on the conventional loans represent confidence intervals. Because the data from the SSBF are from a probability survey based on random selections, this sample is only one of a large number of samples that might have been drawn. Since each sample could have provided different estimates, we express our confidence in the precision of the particular results as a 95 percent confidence interval. This is the interval that would contain the actual population value for 95 percent of the samples that could have been drawn. As a result, we are 95 percent confident that each of the confidence intervals in this report will include the true values in the study population. Data on SBA 7(a) loans do not have confidence intervals because we obtained data on all the loans SBA approved and disbursed from 2001 to 2004.

Figure 2. Percentage of 7(a) and Conventional Loans by Minority Status of Ownership, 2001-2004.

Compared with conventional loans, a higher percentage of 7(a) loans went to small start-up firms from 2001 through 2004 (figure 3).[29]

Specifically, 25 percent of 7(a) loans went to small business start-ups from 2001 through 2004. In contrast, an estimated 5 percent of conventional loans went to newer small businesses over the same period.

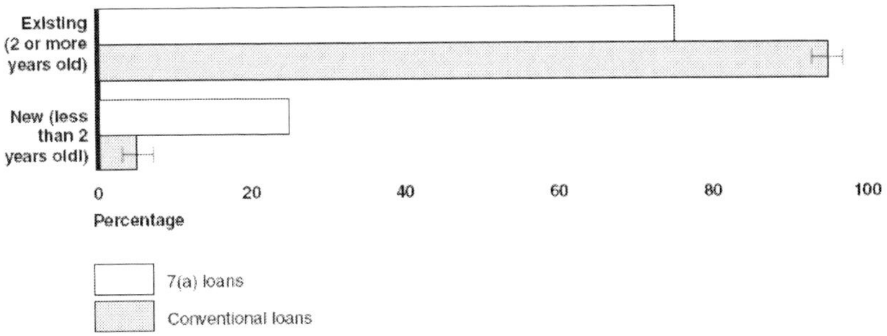

Source: GAO analysis of SBA and Federal Reserve Board of Governors' data.
Note: The brackets on the conventional loans represent a 95 percent confidence interval.

Figure 3. Percentage of 7(a) and Conventional Loans by Status as a New Business, 2001-2004.

MORE SIMILAR PROPORTIONS OF 7(A) AND CONVENTIONAL LOANS SERVED OTHER SEGMENTS OF THE SMALL BUSINESS LENDING MARKET

Compared with the differences in the shares of 7(a) and conventional loans going to minority-owned and start-up small businesses, only limited differences exist between the shares of 7(a) and conventional loans that went to other types of small businesses from 2001 through 2004. For example, the share of 7(a) loans going to small women-owned firms was much closer to the estimated share of conventional loans going to these firms. Specifically, women-owned firms received 22 percent of all 7(a) loans and an estimated 16 percent of conventional loans (figure 4). Furthermore, the percentages of loans going to firms owned equally by men and women were also more similar—17 percent of 7(a) loans and an estimated 14 percent of conventional loans (see figure 4). However, these percentages are small compared with those for small firms headed by men, which captured most of the small business lending market from 2001 to 2004. These small businesses received an estimated 70 percent of conventional loans and 61 percent of 7(a) loans.

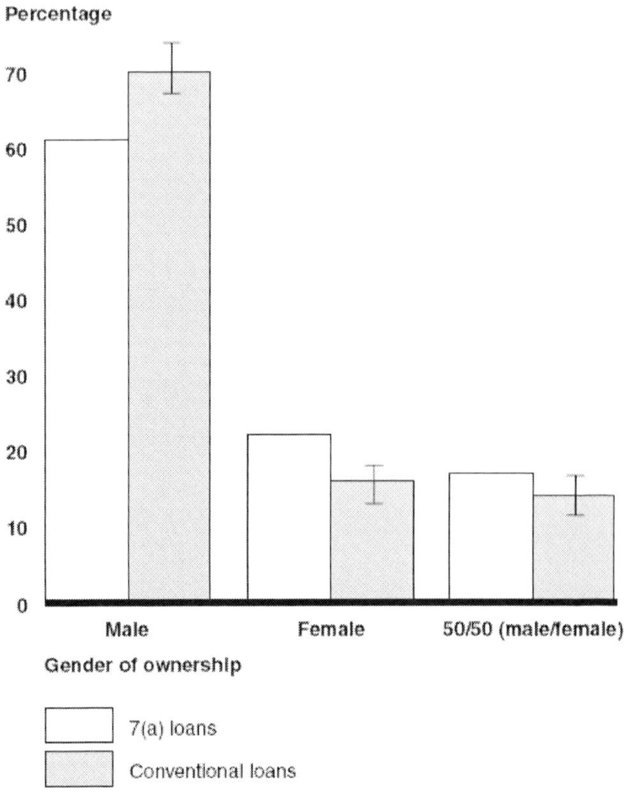

Source: GAO analysis of SBA and Federal Reserve Board of Governors' data.
Note: The brackets on the conventional loans represent a 95 percent confidence interval.

Figure 4. Percentage of 7(a) and Conventional Loans by Gender of Ownership, 2001-2004.

Similarly, compared with the differences in the shares of 7(a) and conventional loans going to minority-owned and start-up small businesses, relatively equal shares of 7(a) and conventional loans reached small businesses in economically distressed neighborhoods (i.e., zip code areas) from 2001 through 2004—14 percent of 7(a) loans and an estimated 10 percent of conventional loans.[30] In order to apply a single measure uniformly across the country, we based our measure on the minimum poverty level eligibility requirement employed by two federal programs designed to assist distressed communities.[31] Specifically, we defined distressed neighborhoods as zip code areas where at least 20 percent of the population had incomes below the national poverty line (see app. I for more information on our methodology).

SBA does not specifically report whether a firm uses its 7(a) loan in an economically distressed neighborhood. Nevertheless, SBA does track loans that go to firms located in areas it considers "underserved" by the conventional lending market. SBA defines an "underserved" area as any one of these federally defined areas: Historically Underutilized Business Zone, Empowerment Zone/Enterprise Community, low- and moderate-income census tract (median income of census tract is no greater than 80 percent of the associated metropolitan area or nonmetropolitan median income), or rural as classified by the U.S. Census.[32] Using this measure, SBA's analysis found that 49 percent of 7(a) loans approved and disbursed in fiscal year 2006 went to geographic areas that SBA considered "underserved" by the conventional lending market.

Although a higher proportion of 7(a) loans went to smaller firms (that is, firms with up to 5 employees), we found that the differences in the shares of 7(a) and conventional loans were more similar for categories of larger firms that have 5 or more employees. Specifically, 57 percent of all 7(a) loans went to small businesses with up to 5 employees, compared with the estimated 42 percent of conventional loans that went to firms with a similar number of employees. In contrast, firms with 5 to 9 employees received 21 percent of the 7(a) loans and 24 percent of conventional loans, and firms with 10 to 19 employees received 12 percent of 7(a) loans and 17 percent of conventional loans. Firms with 20 to 499 employees (the maximum number of employees a business can have and still be considered small by SBA's standards) also received more similar shares of 7(a) and conventional loans.[33]

More similar proportions of 7(a) and conventional loans also went to small businesses with different types of organizational structures and in different geographic locations. For instance, between 2001 and 2004 most 7(a) loans (69 percent) and most conventional loans (an estimated 60 percent) went to corporations.[34] Additionally, similar shares of 7(a) loans (28 percent) and conventional loans (approximately 32 percent) went to sole proprietorships. Similar percentages of 7(a) and conventional loans went to small firms across geographic locations (based on the nine Census divisions). The central regions of the country (e.g., Mountain, West North

Central, and West South Central) received the most similar shares of 7(a) and conventional loans (figure 5).

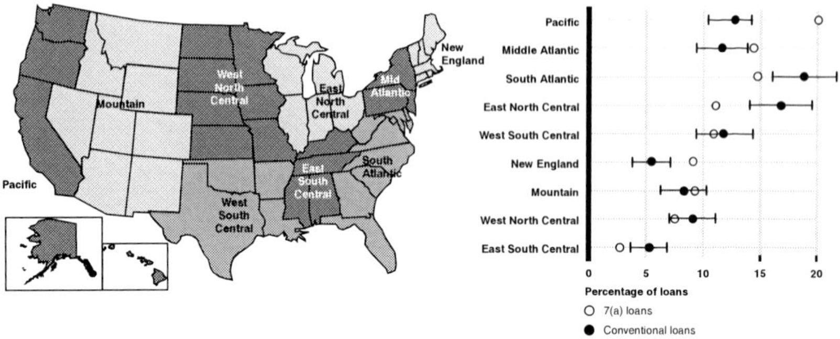

Sources: GAO analysis of SBA data; Art Explosion (map).

Note: The brackets on the conventional loans represent a 95 percent confidence interval.

Figure 5. Percentage of 7(a) and Conventional Loans by Census Divisions, 2001-2004.

Our analysis of information on the credit scores of small businesses that accessed credit without SBA assistance showed only limited differences in these credit scores and those of small firms that received 7(a) loans. As reported in a database developed by two private business research and information providers, The Dun and Bradstreet Corporation and Fair Isaac Corporation (D and B/FIC), the credit scores we compared are typically used to predict the likelihood that a borrower, in this case a small business, will repay a loan.[35] In our comparison of firms that received 7(a) loans and those that received conventional credit, we found that for any particular credit score band (e.g., 160-<170) the differences were no greater than 5 percentage points and the average difference for these credit score bands was 1.7 percentage points (see figure 6). More credit scores for 7(a) borrowers were concentrated in the lowest (i.e., more risky) bands compared with general borrowers, but most firms in both the 7(a) and the D and B/FIC portfolios had credit scores of between 170 and 200. Finally, the percentage of firms that had credit scores in excess of 210 was less than 1 percent for both groups.

The results of our analysis of credit scores should be interpreted with some caution. First, the time periods for the two sets of credit scores are different. Initial credit scores for businesses receiving 7(a) loans in our analysis are from 2003 to 2006.[36] The scores developed by D and B/FIC for small businesses receiving conventional credit are based on data from 1996 through 2000 that include information on outstanding loans that may have originated during or many years before that period.[37] Second, D and B/FIC's scores for small businesses receiving conventional loans may not be representative of the population of small businesses. Although D and B/FIC combined hundreds of thousands of financial records from many lenders and various loan products with consumer credit data for their credit score development sample, they explained that the sample was not statistically representative of all small businesses.

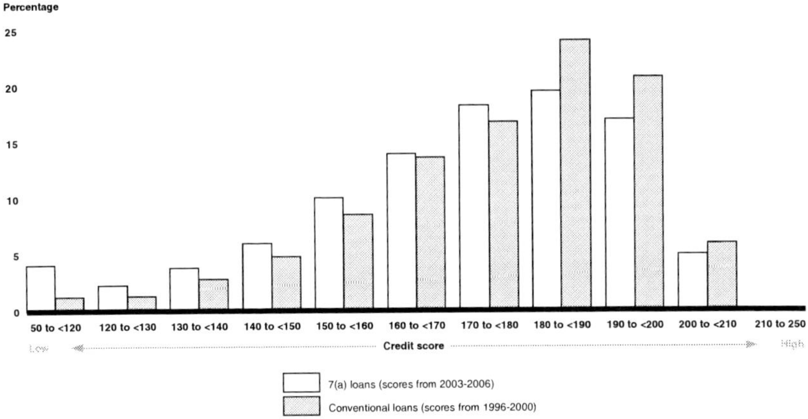

Source: GAO analysis of initial credit scores for loans in the SBA portfolio (2003-2006) and D and B/FIC's analysis of credit scores from data on small businesses in the small business portfolio score (SBPS) development sample (1996-2000).

Note: See app. I for details on the data used to perform this analysis.

Figure 6. Percentage of Small Business Credit Scores (2003-2006) for Firms That Received 7(a) and Conventional Credit in D and B/FIC Sample (1996-2000), by Credit Score Range.

Another score developed by D and B, called the Financial Stress Score (FSS), gauges the likelihood that a firm will experience financial stress—for example, that it will go out of business.[38] SBA officials said that based on analyses of these scores, the difference in the repayment risk of lending associated with 7(a) loans was higher than the risk posed by small firms able to access credit in the conventional lending market. According to an analysis D and B performed based on these scores, 32 percent of 7(a) firms showed a moderate to high risk of ceasing operations with unpaid obligations in 2006, while only 17 percent of general small businesses had a similar risk profile.

7(A) LOANS TENDED TO BE LARGER THAN CONVENTIONAL LOANS AND TO HAVE VARIABLE RATES, LONGER MATURITIES, AND HIGHER INTEREST RATES

Compared with conventional loans, a greater percentage of 7(a) loans were for larger dollar amounts. For example, 61 percent of the number of 7(a) loans had dollar amounts in the range of more than $50,000 to $2 million (the maximum 7(a) loan amount), compared to an estimated 44 percent of the number of conventional loans (see figure 7).[39] A larger share of conventional loans had dollar amounts of $50,000 or less—an estimated 53 percent, compared with 39 percent of 7(a) loans.

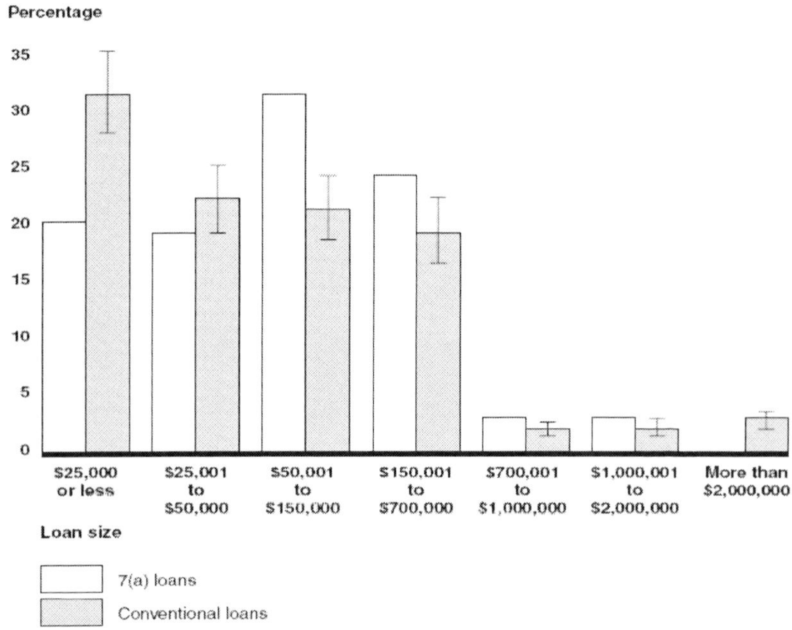

Source: GAO analysis of SBA and Federal Reserve Board of Governors' data.

Note: The brackets on the conventional loans represent a 95 percent confidence interval. The maximum gross 7(a) loan amount is $2 million. The dollar range categories on this chart reflect program thresholds for loan amounts associated with different interest rates or guarantee fee levels.

Figure 7. Percentage of 7(a) Loans and Conventional Loans by Loan Size, 2001-2004.

Although more conventional than 7(a) loans were made for smaller amounts (i.e., less than $50,000), a higher proportion of conventional loan dollars were concentrated in the highest loan amount category (i.e., more than $2 million). In contrast, 70 percent of loans with 7(a) guarantees were for amounts less than $150,000, while 78 percent of 7(a) loan dollars were concentrated in loans with amounts of $150,000 or greater. In addition, almost all 7(a) loans had variable interest rates and maturities that tended to exceed those for conventional loans. Nearly 90 percent of all 7(a) loans but only an estimated 43 percent of conventional loans had variable rates, and, almost 80 percent of 7(a) loans had maturities of more than 5 years, compared with 5 years or less for an estimated 83 percent of conventional loans (figure 8).

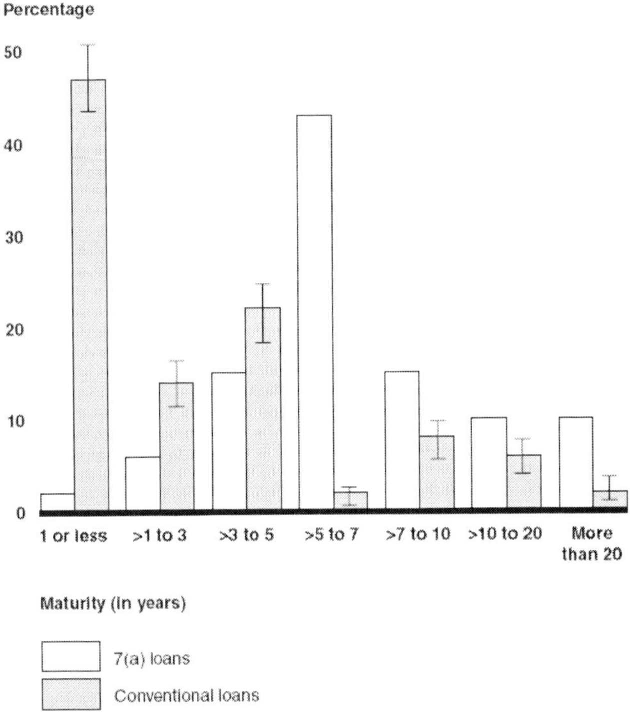

Source: GAO analysis of SBA and Federal Reserve Board of Governors' data.
Note: The brackets on the conventional loans represent a 95 percent confidence interval.

Figure 8. Percentage of 7(a) and Conventional Loans by Loan Maturity Category, 2001-2004.

Finally, for loans under $1 million, interest rates were generally higher for 7(a) loans than for conventional loans. As shown in figure 9, from 2001 through 2004 quarterly interest rates for loans guaranteed by the 7(a) program were on average an estimated 1.8 percentage points higher than interest rates for conventional loans.[40] Interest rates for small business loans offered in the conventional market tracked the prime rate closely and were, on average, an estimated 0.4 percentage points higher.[41] Because the maximum interest rate allowed by the 7(a) program was the prime rate plus 2.25 percent or more, over the period, the quarterly interest rate for 7(a) loans on average exceeded the prime rate.[42]

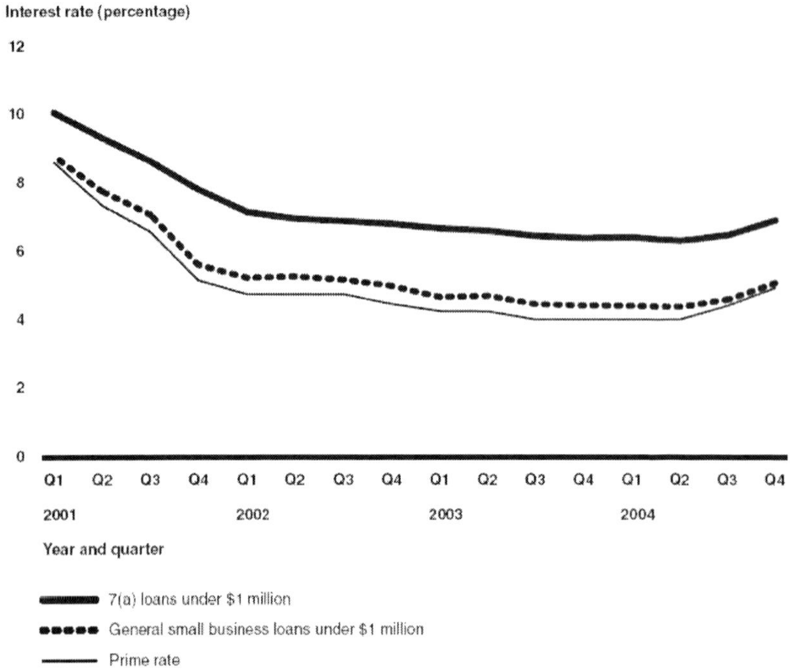

Source: GAO's analysis of SBA data, the Federal Reserve Board of Governors' quarterly *Survey of Terms of Bank Lending* (2001 to 2004), and the Federal Reserve Board of Governors' H.15 statistical release for bank prime loan rates.

Figure 9. Interest Rates Comparison for Loans under $1 Million and Prime Rate, 2001-2004.

Chapter 6

CURRENT REESTIMATES SHOW LOWER-THAN-EXPECTED SUBSIDY COSTS, BUT FINAL COSTS MAY BE HIGHER OR LOWER FOR SEVERAL REASONS

SBA has predicted that the current reestimated credit subsidy costs of 7(a) loans made during fiscal years 1992 through 2004 generally will be lower than the original estimates (see figure 10). Original estimates are made at least a year before any loan is made. The credit subsidy cost is often expressed as a percentage of loan amounts—that is, a credit subsidy rate of 1 percent indicates a subsidy cost of $1 for each $100 of loans. As figure 10 shows, the original credit subsidy cost estimated for fiscal years 2005 and 2006 was zero, since the 7(a) program became a "zero credit subsidy" program. Although the federal budget recognizes costs as loans are made and adjusts for these costs throughout the lives of the loans, the ultimate cost to taxpayers is certain only when none of the loans in a cohort remain outstanding and the agency makes a final, closing reestimate. For loans made in fiscal years 2005 and 2006, SBA adjusted the ongoing servicing fee it charges participating lenders so that the initial subsidy estimate would be zero based on expected loan performance at that time. In addition to the subsidy costs, SBA incurs administrative expenses for operating the loan guarantee program, though these costs are appropriated separately from the cost of the credit subsidy. In its fiscal year 2007 budget request, SBA requested nearly $80 million to cover administrative costs associated with the 7(a) program.

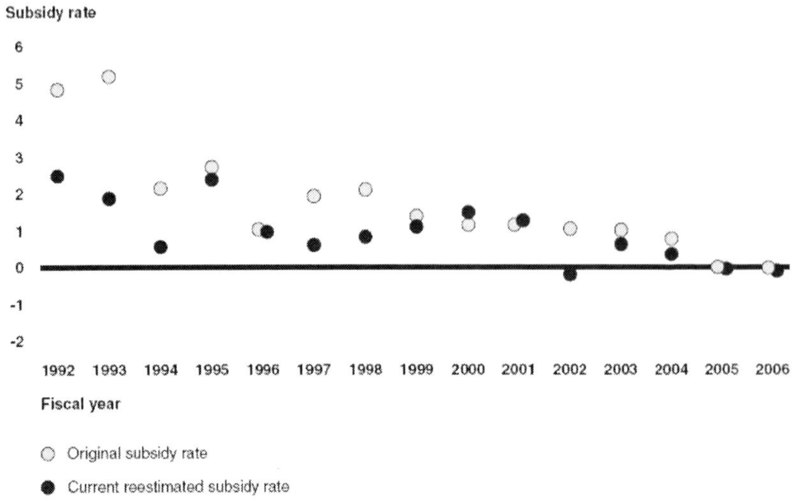

Source: 2008 Federal Credit Supplement.

Figure 10. Original and Current Reestimated Credit Subsidy Rates for Loans Made from 1992 through 2006.

Any forecasts of the expected costs of a loan guarantee program such as 7(a) are subject to change, since the forecasts are unlikely to include all the changes in the factors that can influence the estimates. In part, the estimates are based on predictions about borrowers' behavior—how many borrowers will pay early or late or default on their loans and at what point in time. According to SBA officials, loan defaults are the factor that exerts the most influence on the 7(a) credit subsidy cost estimates and are themselves influenced by various economic factors, such as the prevailing interest rates. Since the 7(a) program primarily provides variable rate loans, changes in the prevailing interest rates would result in higher or lower loan payments, affecting borrowers' ability to pay and subsequently influencing default and prepayment rates. For example, if the prevailing interest rates fall, more firms could prepay their loans to take advantage of lower interest rates, resulting in fewer fees for SBA. Loan defaults could also be affected by changes in the national or a regional economy. Generally, as economic conditions worsen—for example, as unemployment rises—loan defaults increase. To the extent that SBA cannot anticipate these changes in the initial estimates, it would include them in the reestimates.

Beginning in fiscal year 2003, SBA has employed an econometric model that incorporates historical data and other economic assumptions for its credit subsidy cost estimates and reestimates instead of relying primarily on predictions based on historical average loan performance. In previous work we found that the econometric models SBA used to estimate defaults, prepayments, and recoveries were reasonable but that the agency could expand the type of data it used and its method of documenting its decisions regarding the models.[43] According to SBA officials, the agency has made some recent enhancements to the 7(a) credit subsidy cost model, including using more current financial data on borrowers participating in the 7(a) program. SBA officials explained that the agency had also begun validating loan data extracted for use in its econometric model by comparing these data to cohort- and program-level data from another SBA database containing summary loan data. Further, the model now better accounts for amounts SBA recovers from borrowers. SBA officials said that the annual review the agency conducts of the 7(a) credit subsidy cost model may result in minor future changes but that those changes would probably not have any significant impact on the subsidy estimates and reestimates.

Chapter 7

CONCLUSIONS

According to the 7(a) loan program's underlying statutes and legislative history, 7(a) is intended to supplement, not compete with, the conventional lending market by helping address credit constraints that small businesses face. The 7(a) program's design is consistent with this intent—for example, the program's credit elsewhere requirement is designed to help ensure that loans made through the 7(a) program do not supplant credit already available in the conventional lending market. Reflecting the evolving mission of the program, 7(a)'s performance measures focus on the extent to which the program provides guaranteed loans to distinct groups of small businesses, such as start-ups and those whose owners face "special competitive opportunity gaps," including minority- or women-owned businesses. Our evaluation of the program's performance measures found that they were useful in showing how many loans had been made—that is the measures effectively show outputs, but that they did not provide adequate information on the extent to which SBA was meeting its strategic goal of helping small businesses succeed by identifying outcomes. As a result, the actual impact of the 7(a) program remains unclear.

Further, only limited evidence exists on the extent to which small businesses face credit constraints, such as credit rationing, in the conventional lending market. The studies we reviewed suggest that some small firms may face credit rationing within the conventional lending market, but these studies relied on older data. As a result, they did not account for recent trends in the conventional lending market, such as the use of credit scores, that could impact lending to small businesses by providing lenders with additional information to assess a small firm's risk. The effect that these developments may have on the credit constraints that some small businesses face is not yet known.

Based on our analysis, the 7(a) loan program appears to serve certain segments of the small business lending market in different proportions than conventional loans. A higher proportion of 7(a) loans went to minority-owned firms and start-ups, and these results are consistent with the program's legislative intent. But the shares of 7(a) and conventional loans that went to other segments of the small business lending market, such as women-owned businesses and those located in economically distressed areas, were more similar. These results may be useful to SBA as it considers how it administers the program, including its efforts to promote the 7(a) program to lenders and small businesses, and how it oversees participating lenders.

Beginning with fiscal year 2005, the 7(a) program's credit subsidy cost has been estimated at zero; however, the credit subsidy costs estimated for any fiscal year can change due to various factors and are not final until no loans from that year's cohort remain outstanding. Current credit subsidy reestimates of loans made in fiscal years prior to 2005 are lower than originally estimated. Nevertheless, changes in certain important factors, such as 7(a) loan defaults, can influence the 7(a) program's credit subsidy costs.

Recognizing its lack of outcome-based information on the firms that the 7(a) program assists, SBA has efforts underway to develop and implement performance measures to better track outcomes of the 7(a) program including how small firms fare after they participate in the 7(a) loan program. However, SBA has not made clear when, or even if, it plans to complete these efforts, in part because of the costs and legal concerns associated with obtaining the necessary information to undertake this impact analysis. Furthermore, since firms with SBA-guaranteed loans represent various geographic areas, go to both existing and new businesses, and have loan terms sensitive to prevailing economic conditions, many factors unrelated to the loans may impact how well firms do after receiving assistance. It is also unclear what benchmark for success SBA should adopt for these firms. But without some information on outcomes, SBA is unable to provide clear evidence about the impact its 7(a) program is having on firms it assists.

Firms able to meet their loan obligations signal that their businesses are continuing to operate in the communities they are located in and are, at a minimum, experiencing enough financial success to repay their loans. SBA already has loan performance data, such as the number of loans that are in default, prepaid, or in good standing, and other information on firms that receive assistance from the 7(a) program. These data may be reasonable proxies for how well firms are faring after receiving guaranteed loans. In addition, although SBA could incur costs for collecting additional outcome-based information, data reflecting the success of assisted businesses—such as the number that go out of business or begin to rely on conventional credit—could be useful performance measures.

Chapter 8

RECOMMENDATION FOR EXECUTIVE ACTION

To better ensure that the 7(a) program is meeting its mission responsibility of helping small firms succeed through guaranteed loans, we recommend that the SBA Administrator complete and expand SBA's current work on evaluating the program's performance measures. As part of this effort, at a minimum SBA should further utilize the loan performance information it already collects, including but not limited to defaults, prepayments, and number of loans in good standing, to better report how small businesses fare after they participate in the 7(a) program.

Appendix I: Objectives, Scope and Methodology

In this report, we examined (1) the statutory framework and legislative history of the 7(a) program and performance measures the Small Business Administration (SBA) utilizes to assess program results; (2) factors in the conventional lending market that may affect small businesses' access to credit, including market imperfections; (3) how the segments of the small business lending market served by 7(a) loans compare with segments served by conventional loans; and (4) differences in SBA's estimates and reestimates of 7(a)'s credit subsidy costs and the factors that may cause uncertainty about the costs of the 7(a) program to the federal government.

Analysis of Statutory Framework of 7(a) Program and Its Performance Measures

To describe the purpose of the 7(a) program, we reviewed the program's underlying statutes and legislative history to understand how the program was intended to help small businesses. To assess SBA's performance measures for the 7(a) program, we selected performance measures specific to the 7(a) program as reported in the SBA's recent Performance and Accountability Reports. We evaluated nine different performance measures against six attributes identified in our earlier work as being indicative of successful performance measures.

Taken from SBA's fiscal year 2006 Performance and Accountability Report, the nine performance measures were:

number of new loans approved to start-up small businesses,

number of new loans funded to start-up small businesses,

number of start-up small businesses assisted,

number of new loans approved to existing small businesses,

number of new loans funded to existing small businesses,

number of existing small businesses assisted,

number of new loans approved to small businesses facing special competitive opportunity gaps,

number of new loans funded to small businesses facing special competitive opportunity gaps, and

number of small businesses facing special competitive opportunity gaps assisted.

Taken from our earlier report, *Tax Administration: IRS Needs to Further Refine Its Tax Filing Season Performance Measures* (GAO-03-143), the six attributes we assessed the above mentioned performance measures against were:

core program activity (measures cover the activities that an entity is expected to perform to support the intent of the program),

measurable target (measure has a numerical goal),

reliability (measure produces the same result under similar conditions),

clarity (measure is clearly stated and the name and definition are consistent with the methodology used to calculate it),

objectivity (measure is reasonably free from significant bias or manipulation), and

linkage (measure is aligned with division and agencywide goals and mission).

We reviewed and summarized agency documents relating to its ongoing contract with the Urban Institute regarding evaluative studies of SBA's lending programs, including the 7(a) program, currently underway. We also interviewed SBA officials to understand agency efforts to improve its 7(a) program performance measures.

ECONOMIC LITERATURE ON CREDIT RATIONING AND DISCRIMINATION

To identify constraints that may limit credit to small businesses we summarized published, peer-reviewed articles that discuss the subject of credit rationing with regard to firms. We identified articles through reviews of citations of the most recent literature, and by identifying current papers that cite the influential papers in this field (e.g., Stiglitz and Weiss (1981)), and by using article search engines, such as "google scholar" and "jstor." The review concentrated on empirical studies of the U.S. financial market, although studies of the non-U.S. market were included in order to understand the various empirical methodologies employed in this area. In addition, we summarized recent peer-reviewed studies that explore the extent of racial, ethnic, and gender disparities within the conventional lending market. Studies published by think tanks and others that were not peer-reviewed were not included in our review. Appendix II includes a more detailed description of the studies we reviewed about credit rationing and discrimination.

COMPARISON BETWEEN 7(A) AND CONVENTIONAL LOANS

As described more fully in the following sections, to assess similarities and differences in the small business lending market segments served by 7(a) and conventional loans, we compared relevant information on loan terms and borrower characteristics using several data sources. Our analysis was restricted to loans made to firms located within the 50 states, and did not include Puerto Rico or any U.S. territories. To assess the reliability of the data used, we reviewed applicable documentation associated with the specific data source, such as a data dictionary, survey questionnaire, and methodology report. We interviewed officials at the Board of Governors of the Federal Reserve System (Federal Reserve) and SBA who provided us the data in order to understand any data limitations and how the data are collected and stored. We also consulted with Dun and Bradstreet Corporation and Fair Isaac Corporation (D and B/FIC) officials about their data used to generate credit scores for small businesses, including those used by the SBA. Finally, we conducted logic and electronic tests of each data source. We determined the data to be sufficiently reliable for use in our report.

NUMBER OF LOANS AND LOAN DOLLARS OUTSTANDING

To compare the number and amount of outstanding small business loans to 7(a) loans, we used the Federal Deposit Insurance Corporation's (FDIC) Consolidated Reports of Condition and Income (call reports) for U.S. banks. U.S. commercial banks and insured savings institutions are required by federal law to report certain financial information to their appropriate bank regulator quarterly, which FDIC then consolidates and maintains in a database. For the purposes of the call reports, a small business loan is defined by SBA's Office of Advocacy as a commercial and industrial loan or a non-farm, nonresidential loan for which the original amount was $1 million or less. Therefore, we considered the call report data on loans under $1 million to be a proxy for general small business loans, even though there is no attempt to directly link the loans to the size of the firm accessing credit in the call report data. SBA reports tabulations of call report data prepared for the agency by an external contractor as of June 2005, the latest data available. We requested that SBA provide us with similar information on the number and amount of outstanding 7(a) loans under $1 million as of September 30, 2005.

LOAN AND BORROWER CHARACTERISTICS

To evaluate SBA's 7(a) borrowers and loan terms, we used data from two SBA administrative data systems: (1) the Loan Accounting System and (2) the Loan/Lender Monitoring System for information to describe 7(a) loans and borrowers. To assess general small business borrowers and loan terms, we used the 2003 Survey of Small Business Finances (SSBF) conducted by the Federal Reserve. We also used Federal Reserve's historical reports on the monthly bank prime rate in its *Survey of Terms of Bank Lending* to report the quarterly interest rates for loans under $1 million. In addition, we obtained from the D and B/FIC small business credit scores derived from their Small Business Predictive Score development sample.

SBA's data include various information describing the loan, such as the percentage of the loan guaranteed by SBA, the number of months to maturity, and whether the loan had a fixed or variable interest rate. The data also include information on the small firm, such as the ethnicity and gender of the principal owner, the number of employees, and the firm's status as new (i.e., less than 2 years old). SBA provided us with 304,032 records from its administrative data systems, which contained information on all loans approved and disbursed in calendar years 2001 through 2005. Based on discussions with SBA officials about the data and logic testing, we eliminated certain cases from the data provided that had missing values, zero values where appropriate, or that SBA officials confirmed as incorrect data. We eliminated records with any missing or confirmed incorrect information in order to have the same number of cases for each analysis performed. This reduced the number of 7(a) records by 7,495.[1] SBA officials identified an additional 1,730 incorrect social security numbers, which further reduced the number of 7(a) records. We also eliminated 24,010 records to delete multiple loans to the same business. In order to make the SBA data more comparable to the SSBF data, we included only SBA loans a borrower received between 2001 and 2004, which further reduced the number of 7(a) records by 78,056. The final number of 7(a) records we used in our analysis was 192,741, representing a 36 percent decrease in the number of records originally provided by SBA.

We used information from the SSBF as a proxy for loans made to small firms within the conventional lending market (i.e., not made with the assistance of the 7(a) program).[2] The SSBF interviewed 4,240 firms in 2004 and early 2005 that were selected to provide a representative sample of all small businesses in the United States.[3] Among other things, the SSBF assesses credit availability for small businesses, provides financial data for small businesses currently unavailable from other sources, and validates geographic and product market definitions. SSBF data are used to study the effects of changes within the lending industry on credit use by small businesses and to monitor technological and competitive changes in markets for financial services used by small businesses. We used records in which firms reported that the last loan they had applied for had been approved. Applying this standard reduced the number of records by 2,479. We further eliminated records in which firms reported obtaining their most recent loan outside of 2001 to 2004 and firms reporting zero employees, which further decreased the number of records by 23. The final unweighted number of records from the SSBF data was 1,738. Since the data were from a sample with statistical weights, all the percentages in the body of the report reflect weighted percentages. In addition, the SSBF includes multiple imputed values. Our standard error and confidence interval calculations incorporate the multiple imputations where appropriate. We calculated the standard error and confidence intervals for each of the analyses performed using these data since they are based on a random sample. Unless otherwise noted, all percentage estimates have a 95 percent confidence interval within plus or minus 5 percentage points.

The following are more detailed descriptions of actions we took to make the data from SBA and SSBF more comparable:

Minority Status of Ownership

SBA's data include an indicator for whether more than 50 percent of the small business owners are from racial or ethnic minority categories. For the first time, the 2003 SSBF combined data on up to three owners and calculated various indicators by majority owner share. The SSBF data included two data fields related to race and ethnicity that we used. The first field designated whether more than 50 percent of the ownership was white, and the second field designated whether 50 percent or more of the ownership was minority or Hispanic. Using these fields, we compared the share of 7(a) and conventional loans that went to small businesses with 50 percent or greater minority ownership.

Longevity of Business

SBA's data include information indicating whether or not the business was new, which SBA defines as being less than 2 years old. The SSBF's information included information on the year of the survey and the year when the firm applied for its most recently approved loan. In addition, the survey included an age for the firm. We calculated the age of the firm when it applied for the most recent loan. We considered a business as new if its age was 2 years or less when it applied for its most recently approved loan.

Number of Employees

The number of employees in SBA's data is the number provided by the prospective borrower at the time of loan approval. According to SBA, the number of employees is required as part of the application process, so any zeros in this field should be treated as missing values. Additionally, we eliminated SBA records that listed the number of employees as 500 or greater to match the SSBF's selection criteria. The SSBF's data included information on the number of full- and part-time employees. All cases specifying zero employees were eliminated.

Gender of Ownership

Both SBA's data and the SSBF's data had information designating whether more than 50 percent, less than 50 percent, or exactly 50 percent of the firm was female-owned. We compared the groups of more than 50 percent female ownership, exactly 50 percent female/male ownership, and more than 50 percent male ownership receiving 7(a) and conventional loans.

Economically Distressed Areas

We created a variable indicating whether or not a given geographic location in which a business receiving a loan is situated, is in economic distress. The indicator we chose was based on the minimum eligibility criteria for the Empowerment Zone and Enterprise Community (EZ/EC) and the Renewal Community (RC) programs, which target federal grant monies to public and private entities, tax benefits to businesses, or both in order to improve conditions in competitively selected, economically distressed communities. The minimum poverty level eligibility requirement for EZ/EC and RC is that at least 20 percent of the population in the census tracts that make up the zone must have incomes below the national poverty line. Using data from the 2000 Census, we used the Census Zip Code Tabulation Areas (which approximate zip code boundaries) to identify zip codes in which 20 percent or more of the individuals had income below the poverty level. We matched the zip codes of businesses receiving 7(a) loans from 2001 through 2004 (using updated geography to account for changes to zip code boundaries) to the 2000 Census file to quantify how many 7(a) loans went to businesses in economically distressed areas. The business locations for respondents to the SSBF are not included in the public use data file. However, Federal Reserve staff matched our distress indicator to the zip codes for their respondents and returned the data to us for merging with the public file without revealing respondents' business locations. We then compared the shares of 7(a) and conventional loans that went to economically distressed areas.

Business Organization

SBA's data included three organizational types—individual (or sole proprietorship), partnership, and corporation. The SSBF included nine organization types—sole proprietorship, partnership, S corporation, C corporation, limited liability partnerships tax filed as partnerships or corporations, and limited liability corporations tax filed as sole proprietorships, partnerships, or corporations. We combined the two types of sole proprietorships, the three types of partnerships, and the four types of corporations in the SSBF's data to provide comparable information.

Geographic Information

The only geographic information in the SSBF's data was the census region in which the firm was located.[4] The state listed in SBA's data was used to group the 7(a) data according to census regions.

Interest Rates

In order to compare interest rates on 7(a) loans to loans general small businesses obtained in the conventional lending market, we used data from the Federal Reserve's *Survey of Terms of Bank Lending*. The survey provides information quarterly on the number of commercial and industrial loans by four size categories (less than $100,000; between $100,000 and $999,999; between $1 million and $999,999,000; and $10 million or more) made only by commercial banks.[5] The survey reports an average interest rate in each category that is weighted by loan amount. We only used data related to the first two categories because those loan amounts most resembled the 7(a) loans in the SBA data and because SBA's Office of Advocacy considers in call report data, discussed previously, loans of $1 million or less to be for small businesses. Limitations to these data regarding our analysis include that the information is gathered during 1 week in the middle month of each quarter and does not distinguish between the sizes of the business obtaining the loan.[6] In addition, the data in the survey do not include loans made by finance companies or small business loans made on credit cards. In order to compare interest rate data we derived from the survey, we used SBA data to calculate the calendar year and quarter in which each loan was disbursed and calculated the average interest rates for all loans disbursed in a given quarter that were for under $1 million. In order to be consistent with the survey, we calculated the average quarterly interest rate using the loan amounts as weights. Finally, we used Federal Reserve's historical reports on the monthly bank prime rate to estimate the prime rate for every quarter from 2001 through 2004.

Credit Scores

To assess the relative creditworthiness of firms receiving 7(a) loans to firms receiving conventional credit, we compared the initial credit scores for loans in SBA's 7(a) portfolio to scores calculated from D and B/FIC's large sample of data from small businesses in the conventional lending market and from consumer credit bureaus. In comparing credit scores for 7(a) firms with other firms, we relied on D and B/FIC's analysis of credit scores based on data from small business transactions, consumer credit bureaus, and loan performance from their user's lending portfolios from 1996 through 2000, known as the Small Business Predictive Score (SBPS) development sample. The loans D and B/FIC used for its sample were outstanding loans including those that originated between 1996 and 2000 and older loans. We relied on the D and B/FIC data from a different time period because time and resource constraints prohibited obtaining more recent data. As stated previously, our comparison of credit scores should be interpreted with caution because the data come from different time periods and the D and B/FIC credit scores may not be representative of the population of general small businesses. However, although the data D and B/FIC used to develop its small business credit score may not be statistically representative of all small businesses, the data sample is very large and reflects the broadest and most recent information readily available to us on small business credit scores in the conventional lending market.

Description of Credit Subsidy Cost Estimates and Reestimates

To describe 7(a)'s credit subsidy cost estimates and reestimates we compared SBA's original credit subsidy cost estimates for fiscal years 1992 through 2006 to SBA's reestimates in fiscal year 2008, as reported in the fiscal year 2008 Federal Credit Supplement. We reviewed documents related to the 7(a) credit subsidy cost model, which the agency uses to generate its estimates and reestimates. We also interviewed SBA officials to understand any differences in the reported original credit subsidy cost estimates and subsequent reestimates, as well as to describe what factors may influence future reestimates.

Analysis of 504 Loan Program

We were unable to undertake a similar comparative analysis between 504 loans and loans made to general small businesses within the conventional lending market primarily due to the limited number of observations of conventional loans that were comparable to loans with 504 guarantees and lack of generalizability with the SSBF data. We have included in appendix III information on the characteristics of borrowers and loans financed under SBA's 504 program based on analysis done using data provided by SBA. We performed the same eliminations of observations for missing or incorrect data that we applied to the 7(a) data as described above, which resulted in a 28 percent (from 28,341 to 20,289) decrease in the number of cases used in our analysis.

We performed our work in Washington, D.C., and Chicago from May 2006 through July 2007 in accordance with generally accepted government auditing standards.

Appendix II: Summary of Economic Literature on the Empirical Evidence for Credit Rationing and Discrimination in the Conventional Lending Market

Study	Objective	Data	Method	Conclusions and limitations
Berger, Allen N., and Gregory F. Udell, “Some Evidence on the Empirical Significance of Credit Rationing,” The Journal of Political Economy, vol. 100, no. 5 (1992): 1047–1077.	One implication of credit rationing is that commercial loan rates do not respond quickly to changes in the market interest rate—i.e., are sticky. Objective was to develop and implement a series of empirical tests able to determine whether loan rate stickiness is explained by credit rationing or something else.	Contract information from 1977 through 1988 on approximately 1 million bank loan contracts.	Tested for “stickiness” and whether it is mitigated by specific loan contract features, such as commitment or collateral. Because commitment loans act as insurance against rationing, they can be used as a test for whether stickiness stems from credit rationing. Because rationing is more likely when open market interest rates are high, also examines the proportion of loans that are made in commitment agreement and whether it increases with the interest rate.	Found evidence inconsistent with credit rationing. Could not conclude that stickiness stems from credit rationing, since nearly half of the loan rate stickiness occurs with commitment loans. The proportion of loans that were commitment loans decreased during credit market tightness, the direction opposite from that predicted by credit rationing. Concluded that these results did not disprove the existence of credit rationing of commercial bank borrowers but indicated that rationing does not constitute an important macroeconomic phenomenon.

(Continued)

Study	Objective	Data	Method	Conclusions and limitations
Berger, Allen N., and Gregory F. Udell, “Relationship Lending and Lines of Credit in Small Firm Finance,” *The Journal of Business*, vol. 68, no. 3 (1995): 351–381.	To examine the effect of relationship lending in small firm finance. Hypothesized that banks may acquire private information over the course of a relationship; therefore, focused on lines of credit issued to small business.	1988-89 National Survey of Small Business Finances survey of 3,404 businesses.	Assessed the empirical relationship between relationship lending and collateral. Focused exclusively on lines of credit, using the firm’s age and the number of years it had done business with the lender as measures of how information can change the terms of credit.	Found evidence consistent with credit rationing. Highlighted the role of relationship lending in loan contracts and provided support for credit rationing. The evidence indicated that small firms with longer relationships pay lower interest rates and are also less likely to pledge collateral. Results suggested that banks accumulate increasing amounts of private information over the duration of the bank-borrower relationship and use the information when defining contract terms. Found that results were consistent with theoretical arguments that relationship lending generates valuable information about borrower quality, which is consistent with credit rationing.

(Continued)

Study	Objective	Data	Method	Conclusions and limitations
Berkowitz, Jeremy, and Michelle J. White, "Bankruptcy and Small Firms' Access to Credit," The RAND Journal of *Economics*, vol. 35, no.1 (2004): 69–84.	To examine how personal bankruptcy law affects small firm access to credit by exploiting state variation in assets shielded from bankruptcy proceedings. Because many small business loans are secured with personal credit, hypothesized that firms in high-exemption states are more likely to be denied credit or be credit rationed.	1993 National Survey of Small Business Finances survey of 5,356 small businesses operating as of year-end 1992.	Tested for credit rationing by exploiting state variation in the type and amount of assets shielded from bankruptcy proceedings. This follows from the study's model, derived from economic theory, which suggests that the more assets shielded from bankruptcy, the greater the incentive to declare bankruptcy.	Found evidence of credit rationing but under a broader definition than other studies. According to the study's definition, managers who are denied credit or discouraged from applying have been "credit rationed." Concluded that higher personal exemptions increase credit rationing. Firms are more likely to be denied credit and, if offered credit, at higher interest rates.

(Continued)

Study	Objective	Data	Method	Conclusions and limitations
Blanchflower, David G., Philip B. Levine, and David J. Zimmerman, "Discrimination in the Small-Business Credit Market," The Review of Economics *and Statistics*, vol. 85, no. 4 (2003): 930–943.	To examine the presence of discrimination in the small business credit market.	1993 and 1998 editions of the National Survey of Small Business Finances.	Using a regression approach, tested whether differences in rates of loan denial or interest by demographic group can be explained by differences in credit worthiness or other factors, including credit scores.	Found mixed results with respect to discrimination. Using a large amount of controls, found significant evidence that African American-owned firms face obstacles in obtaining credit, with lower application rates and higher denial rates. Also found that African American-owned firms were charged higher interest rates. The study referred to the magnitude of the difference for African Americans as substantial but could not find evidence of similar discrimination against women or other ethnic groups.

(Continued)

Study	Objective	Data	Method	Conclusions and limitations
Bodt, Eric de, Frederic Lobez, and Jean- Christophe Statnik, "Credit Rationing, Customer Relationship, and the Number of Banks: An Empirical Analysis," European Financial *Management*, vol. 11, no. 2 (2005): 195–228.	To estimate the effect of bank mergers on access to credit.	Data from a questionnaire sent to 4,932 Belgian firms that met certain selection criteria on data quality and being a small business.	Analyzed the relationship between the numbers of banks used by the firm, customer relationship, and credit rationing for these businesses.	Found no general rule that related the number of banks a firm does business with to the extent of credit rationing. For example, found that smaller firms dealing with big main banks should increase the number of banks in order to minimize the probability of being rationed. Larger firms, dealing with local banks, in contrast, should concentrate financing to limit rationing.
Cavalluzzo, Ken S., and Linda C. Cavalluzzo, "Market Structure and Discrimination: The Case of Small Business," *Journal of* Money, Credit and Banking, vol. 30, no. 4 (1998): 771-792.	To estimate the prevalence of prejudicial discrimination in small business lending.	1988-89 National Survey of Small Business Finances survey of 3,404 businesses, including information on applications for credit and their outcome.	Using the insight that the more competitive a market is, the less the likelihood is of prejudicial discrimination, the study regressed interest rates, rates of application, and denial of credit on measures of concentration of the banking industry where loans were made.	Evidence on discrimination was mixed. Found evidence of prejudicial discrimination against Hispanics and Asians. Found that African American-owned small businesses hold fewer loans but did not find that this stemmed from prejudicial treatment. Found that prejudicial discrimination may favor women.

(Continued)

Study	Objective	Data	Method	Conclusions and limitations
Cavalluzzo, Ken S., Linda C. Cavalluzzo, and John D. Wolken, "Competition, Small Business Financing, and Discrimination: Evidence from a New Survey," Journal of *Business*, vol. 75, no. 4 (2002): 641–679.	To examine whether differences in interest rates, rates of denial, and application rates by gender and race can be linked to discrimination.	1993 National Survey of Small Business Finances survey of 5,356 small businesses operating as of year-end 1992.	Using a regression, used bank concentration to identify prejudicial discrimination. Examined loan application, denial, and interest rates, as well as firms discouraged from applying for credit. Used a rich set of control variables, such as whether borrowers had experienced bankruptcy and the borrowers' credit scores.	Found evidence of discrimination in the small business lending market. Found some evidence of prejudicial discrimination against African Americans and more robust evidence of prejudicial discrimination against women.
Cole, Rebel A., "The Importance of Relationships to the Availability of Credit," Journal of Banking and *Finance*, vol. 22 (1998): 959–977.	To examine the effect of preexisting relationships between lenders and firms on credit availability.	1993 National Survey of Small Business Finances survey of 5,356 small businesses operating as of year-end 1992.	Estimated the effect of relationships on credit availability. Used other types of bank services the firms used, as well as length of relationships, as measures of the strength of the relationship. Whether a firm was extended credit was a measure of credit availability.	Provided evidence consistent with credit rationing, concluding that a preexisting relationship between firm and lender increases the chances that credit will be extended but that the length of the relationship is unimportant.

(Continued)

Study	Objective	Data	Method	Conclusions and limitations
Cowling, Marc, and Peter Mitchell, "Is the Small Firms Loan Guarantee Scheme Hazardous for Banks or Helpful to Small Business?" Small Business Economics, vol. 21, no. 1 (2003): 63–71.	To test an underpinning of credit rationing—that the rate of default increases with the cost of capital—i.e., the interest rate.	Data on 42,316 loans issued with collateral provided by the U.K. Small Firm Loan Guarantee Scheme.	Presented two alternative tests. First, estimated the effect of firm and loan level characteristics on default, and second tested for the effect of factors that change over time.	Found that consistent with credit rationing, default rate increases with the interest rate. However, also found that a series of other factors not addressed by the credit rationing literature, such as the loan's purpose, also affect default rate.
Cressy, Robert, "Are Business Startups Debt-Rationed?" The Economic *Journal*, vol. 106 (1996): 1253–1270.	If financial capital affects business survival, this is evidence that credit constraints exist for some businesses. Objective was to examine whether human capital (such as education) might be an alternative explanation. If human capital is correlated with access to credit, then previous studies that failed to correct for this might incorrectly associate financial assets with business survival.	A sample of 2,000 U.K. start-ups that opened business accounts in 1988.	Tested for debt rationing after correcting for human capital. Used several measures for human capital— proprietors' age, education, work experience in the area of the start-up.	Found no evidence for debt rationing. Evidence suggested that human capital is the true determinant of survival and that the importance of financial capital is spurious. Firms with more human capital are more likely to accept a bank's offer. Concluded that, rather than a bank's selecting firms, they self-select for finance and those firms with more human capital are more likely to accept the bank's offer.

(Continued)

Study	Objective	Data	Method	Conclusions and limitations
Holtz-Eakin, Douglas, David Joulfaian, and Harvey S. Rosen, "Sticking It Out: Entrepreneurial Survival and Liquidity Constraints," The Journal of Political *Economy*, vol. 102, no. 1 (1994): 53–75.	To examine why some individuals survive as entrepreneurs and some do not. Focused on the role of access to capital. Tested an implication of credit rationing—that individuals will face liquidity constraints.	1981 and 1985 federal tax return data on individuals who received inheritances.	Tested whether an inheritance affects business survival. One implication of liquidity constraints would be that entrepreneurs who have access to financial resources independent of the credit market, such as inheritances, are more likely to succeed.	Although not on the subject of small business lending, provided support for credit rationing. Results suggested that a sizable inheritance has a small but noticeable effect on business survival and a larger effect on business receipts, which is consistent with an implication of credit rationing and liquidity constraints.
Levenson, Alec R., and Kristen L Willard, "Do Firms Get the Financing They Want? Measuring Credit Rationing Experienced by Small Businesses in the U.S." Small Business *Economics*, vol. 14, no. 2 (2000): 83–94.	To measure the extent to which small businesses in the late 1980s were able to access external credit at a level they desired. The extent to which this is not true forms the upper bound of credit rationing, since some firms denied credit are actually credit unworthy.	1988-89 National Survey of Small Business Finances survey of 3,404 businesses.	To find an upper bound for the existence of credit rationing, estimated the percentage of small businesses denied credit. Included in the analysis firms denied credit and firms discouraged from applying for credit.	Found evidence consistent with credit rationing. Estimated that an upper bound of 6.36% of firms was rationed. The firms that were rationed represented 3.22% and 3.46% of sales and employment in the survey. Consistent with expectations, credit rationing was associated with firm size. While finding evidence consistent with credit rationing, the evidence suggested that equilibrium credit rationing is

(Continued)

Study	Objective	Data	Method	Conclusions and limitations
				economically unimportant for the small firms analyzed.
Perez, Stephen J., "Testing for Credit Rationing: An Application of Disequilibrium Econometrics," *Journal of* Macroeconomics, vol. 20, no. 4 (1998): 721–739.	To test whether firms experience credit rationing by testing for excess demand. If there is no credit rationing, then the market will be at equilibrium and the supply of credit will equal demand.	5,000 firm-year observations from the CompuStat database of publicly traded firms for each year from 1981 through 1991.	Developed a model that allowed an empirical test for credit rationing. To implement the model, used maximum likelihood methods to test three samples of the population: firms with assets less than \$10 million, assets \$10 million to \$25 million, and assets \$25 million to \$50 million.	Concluded that credit rationing exists. In all three samples, concluded that some firms face excess demand and are credit rationed while some do not. Found that the mean probability that the smallest firms are rationed was 61.9%, medium firms 59.1%, largest firms 59.8%. This suggested that smaller firms are more likely to be credit rationed. Did not test for whether the differences were statistically significant.

(Continued)

Study	Objective	Data	Method	Conclusions and limitations
Petersen, Mitchel A., and Raghuram G. Rajan, "The Benefits of Lending Relationships: Evidence from Small Business Data," The Journal of Finance, vol. 49, no. 1 (1994): 3–37.	To test whether ties between a firm and its creditor affect the cost and availability of credit to the firm and whether they mitigate the effect of credit rationing. Argued that "adverse selection and moral hazard may have a sizeable effect when firms are young and small," which made the sample likely to show the effects of credit rationing.	1988-89 National Survey of Small Business Finances survey of 3,404 businesses.	Estimated the effect relationships have on credit availability and interest rates. Using a regression, tested for the significance of a variety of relationship measures, such as relationship length in years, use of other financial services at the bank, and number of other banks the firm borrows from.	Presented evidence consistent with credit rationing. For interest rates, found a small effect of concentrating business with a single bank on the price charged by lenders; found that firms that borrowed from multiple banks had increased interest rates; and that there was little effect on the length of the relationship. On credit availability, found stronger effects of relationships: the availability of credit from institutions increases as the firm spends more time in the financial relationship and increases the number of financial services used, as that concentrates borrowing at that bank. Argued that these results are consistent with credit rationing but might also be consistent with a reduction in lender's expected cost.

(Continued)

Study	Objective	Data	Method	Conclusions and limitations
Sofianos, George, Paul Wachtel, and Arie Melnik, "Loan Commitments and Monetary Policy," *Journal of* Banking and Finance, vol. 14 (1990): 677–689.	To measure the effect of loan commitments on how monetary policy affects the economy. Commitment is an agreement between the bank and the firm to lend an amount but not at a fixed interest rate. Consequently, a loan commitment should, in the short run, prevent a firm from being credit rationed.	A 1973–87 monthly survey of commercial banks conducted by the Board of Governors of the Federal Reserve.	Examined whether loans under commitment are less affected by a period of monetary tightening, since the bank cannot choose to refuse credit.	Presented evidence consistent with credit rationing. While both types of loans are affected by interest rates, found evidence of a differential effect of monetary policy on loans under commitment. Concluded that quantity rationing occurs in the market for bank loans. Also concluded that borrowers' willingness to obtain commitment loans, at an expense, is consistent with the desire to insure against credit rationing.
Trovato, Giovanni, and Marco Alfo, "Credit Rationing and the Financial Structure of Italian Small and Medium Enterprises," Journal of Applied *Economics*, vol. 9, no. 1 (2006): 167–184.	To analyze the effect of credit subsidies on the development of small and medium Italian enterprises.	Survey data from 1989 through 1994 of approximately 1,919 Italian firms.	Tested whether firms that gain subsidies are more likely to reduce their financial constraints and increase investment levels.	Presented evidence consistent with credit rationing. Found that firms' leverage is positively related to the presence of public subsidies.

Source: GAO analysis.

APPENDIX III: DESCRIPTIVE STATISTICS OF 504 LOAN PROGRAM

As stated previously, 504 loans generally provide long-term, fixed-rate financing to small businesses for major fixed assets, such as land and buildings. The following figures provide descriptive statistics for 504 loans approved and disbursed from 2001 through 2004, including information on the characteristics of 504 loans and borrowers. Not all information available for the 7(a) loans described in the body of the report was available for the 504 loans. For example, SBA does not collect interest rate data for 504 loans. Additionally, because 504 loans are only offered with set maturities (mostly 10 or 20 year) and fixed interest rates, there are no data on revolving loans or loans with variable interest rates.

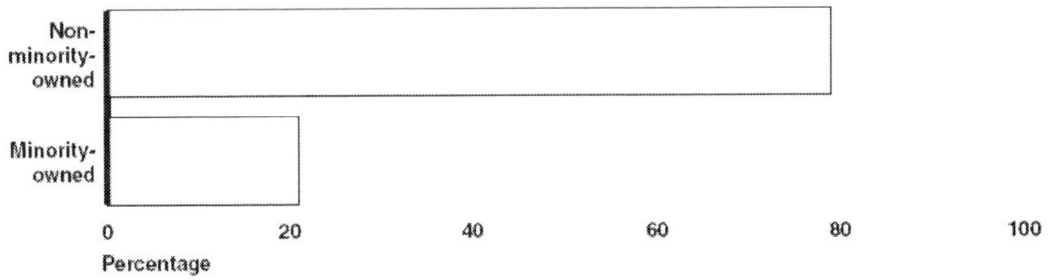

Figure 11. Percentage of 504 Loans by Minority Status of Ownership, 2001-2004.

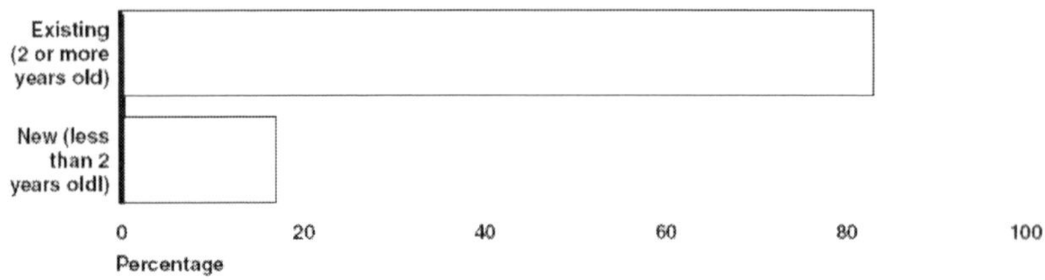

Figure 12. Percentage of 504 Loans by Status as a New Business, 2001-2004.

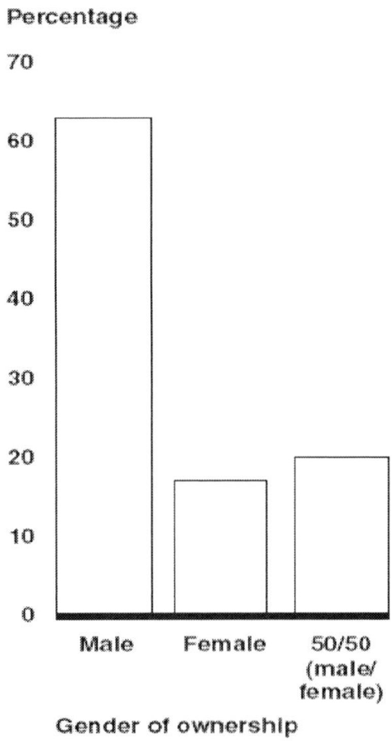

Figure 13. Percentage of 504 Loans by Gender of Ownership, 2001-2004.

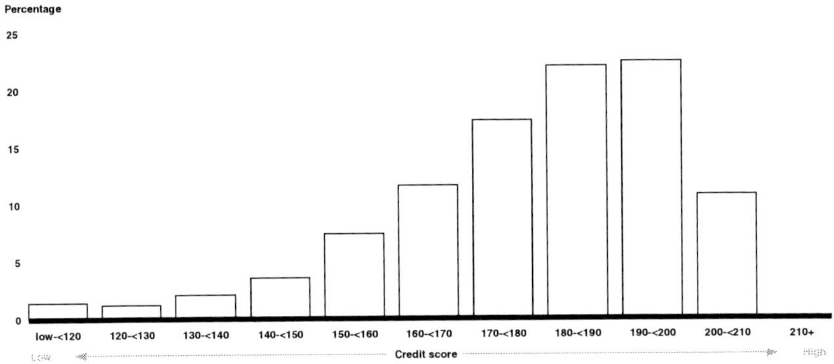

Figure 14. Percentage of Small Business Credit Scores for Firms That Received 504 Loans by Credit Score Range, 2003-2006/

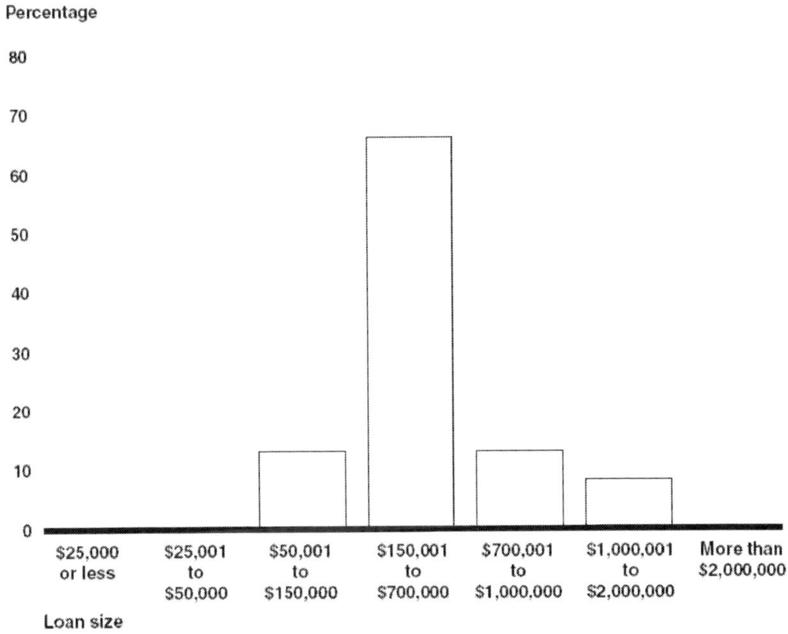

Figure 15. Percentage of 504 Loans by Loan Size, 2001-2004.

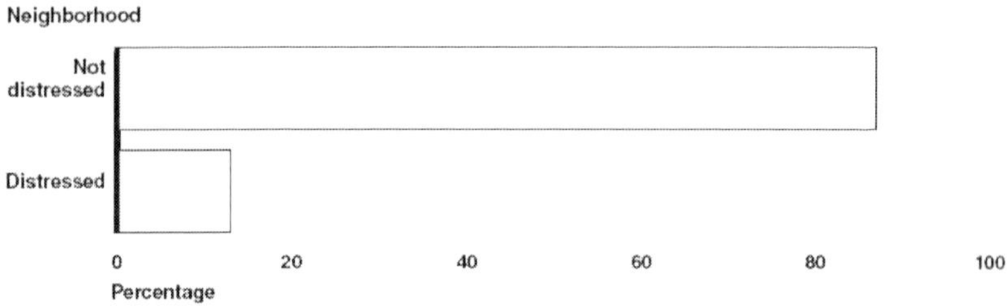

Figure 16. Percentage of 504 Loans in Distressed Neighborhoods, 2001-2004 .

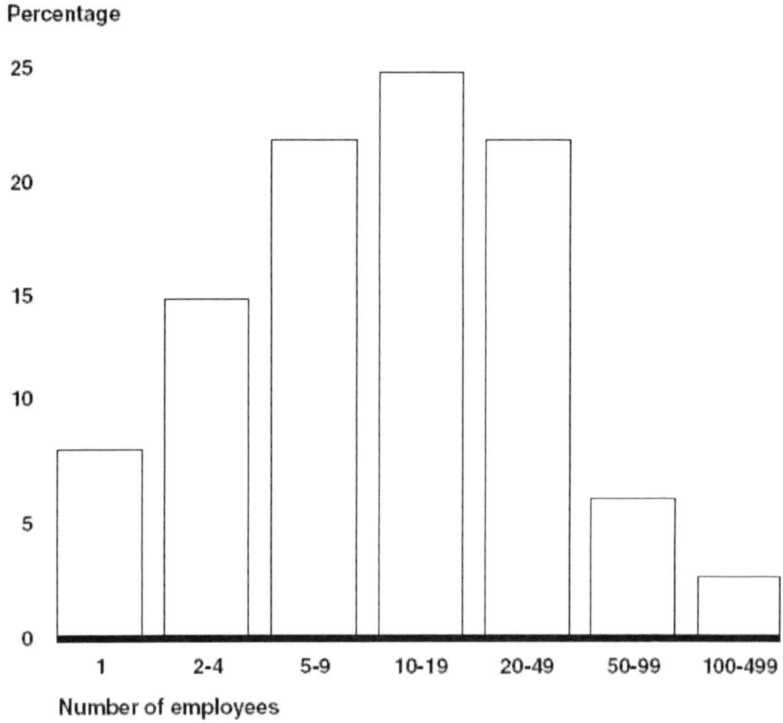

Figure 17. Percentage of 504 Loans by Number of Employees in the Firm, 2001-2004.

Figure 18: Percentage of 504 Loans by Census Divisions, 2001-2004

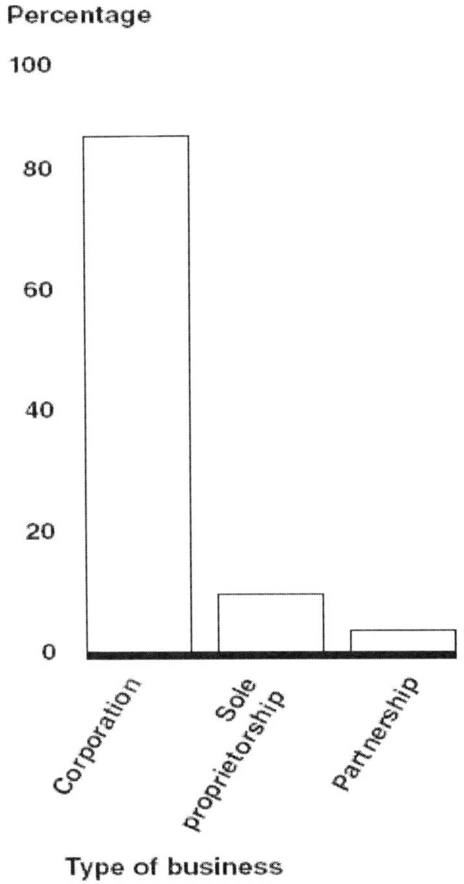

Figure 19. Percentage of 504 Loans by Business Organization Type, 2001-2004.

REFERENCES

[1] Section 7(a) of the Small Business Act, as amended, codified at 15 U.S.C. § 636(a); see also 13 C.F.R. Part 120. Although SBA has limited legislative authority to make direct loans to borrowers unable to obtain loans from conventional lenders, SBA has not received any funding for these programs since fiscal year 1996.

[2] To compare the number and amount of outstanding small business loans to 7(a) loans, we used SBA reports based on the Federal Deposit Insurance Corporation's (FDIC) Consolidated Reports of Condition and Income for U.S. Banks (call reports) and SBA data on outstanding 7(a) loans. In analyzing data from call reports, SBA defines a small business loan as a commercial and industrial loan for which the original amount was less than $1 million.

[3] SBA has data available to make this comparison only for 2003, 2004, and 2005.

[4] Within the 7(a) program, there are several program delivery methods—regular 7(a), the certified lender program, the preferred lender program, SBAExpress, Community Express, Export Express, and Patriot Express. SBA provides final approval for loans made under the regular 7(a) program. Certified lenders must perform a thorough credit analysis on the loan application packages they submit to SBA so that SBA can rely on that analysis to allow it to perform a credit review only, thereby shortening the time for SBA loan processing. Preferred lenders have delegated authority to make SBA-guaranteed loans, subject only to a brief eligibility review and assignment of a loan number by SBA. Lenders participating in

SBAExpress, Community Express, Export Express, and Patriot Express also have delegated authority to make SBA-guaranteed loans.

[5] In establishing size standards, SBA considers economic characteristics comprising the structure of the industry, including degree of competition, average firm size, start-up costs and entry barriers, and distribution of firms by size. It also considers growth trends, competition from other industries, and other factors that may distinguish small firms from other firms. SBA's size standards seek to ensure that a firm that meets a specific size standard is not dominant in its field of operation.

[6] Some earlier work includes GAO, Executive Guide: Effectively Implementing the Government Performance and Results Act, GAO/GGD-96-118 (Washington, D.C.: June 1996) and GAO, The Results Act: An Evaluator's Guide to Assessing Agency Annual Performance Plans ,GAO/GG D-10.1.20 (Washington, D.C.: April 1998).

[7] GAO, Tax Administration: IRS Needs to Further Refine Its Tax Filing Season Performance Measures, GAO-03-143 (Washington, D.C.: Nov. 12, 2002).

[8] GAO, Managing for Results: Opportunities for Continued Improvements in Agencies' Performance Plans, GAO/GGD/AIMD-99-215 (Washington, D.C.: July 20, 1999).

[9] Small Business Administration Inspector General, *Results Act Performance Measurement for the 7(a) Business Loan Program*, Report No. 1-01 (Washington, D.C.: December 2000).

[10] Small business credit scores are a range of numeric values derived using a mathematical model that takes into account information from consumer credit bureaus and business performance data from lenders. The scores attempt to predict the likelihood that a business will repay a loan.

[11] 5 U.S.C. 306(b).

[12] Appendix II identifies and provides information on the studies we reviewed, including their objectives, data, methodologies, limitations, and conclusions.

[13] For more details on how economic theory predicts credit rationing, see J. E. Stiglitz and A. Weiss, "Credit Rationing in Markets with Imperfect Information," *The American Economic Review*, vol. 71, no. 3 (1981).

[14] However, under certain circumstances, economic reasoning suggests that lack of information about certain types of borrowers could result

in the opposite—an excess of credit. See D. De Meza and D.C. Webb, "Too Much Investment: A Problem of Asymmetric Information," *The Quarterly Journal of Economics*, vol. 102, no. 2 (1987).

[15] We also identified additional studies that examined evidence for credit rationing between lenders and borrowers, but these studies were all based on data from foreign countries.

[16] S.J. Perez, "Testing for Credit Rationing: An Application of Disequilibrium Econometrics," *Journal of Macroeconomics*, vol. 20, no. 4 (1998).

[17] A.R. Levison and K.L. Willard, "Do Firms Get the Financing They Want? Measuring Credit Rationing Experienced by Small Businesses in the U.S.," *Small Business Economics*, vol. 14, no. 2 (2000) and A.N. Berger and G.F. Udell, "Some Evidence on the Empirical Significance of Credit Rationing," *The Journal of Political Economy*, vol. 100, no. 5 (1992).

[18] Levinson and Willard, "Do Firms Get the Financing They Want? Measuring Credit Rationing Experienced by Small Businesses in the U.S.," 90.

[19] Berger and Udell, "Some Evidence on the Empirical Significance of Credit Rationing," 1076.

[20] J. Berkowitz and M.J. White, "Bankruptcy and Small Firms' Access to Credit," *The RAND Journal of Economics*, vol. 35, no. 1 (2004).

[21] Levinson and Willard, "Do Firms Get the Financing They Want? Measuring Credit Rationing Experienced by Small Businesses in the U.S.," 90.

[22] M.A. Petersen and R.G. Rajan, "The Benefits of Lending Relationships: Evidence from Small Business Data," *The Journal of Finance*, vol. 49, no. 1 (1994) and R. A. Cole, "The Importance of Relationships to the Availability of Credit," *Journal of Banking and Finance*, vol. 22 (1998).

[23] A.N. Berger and G.F. Udell, "Relationship Lending and Lines of Credit in Small Firm Finance," *The Journal of Business*, vol. 68, no. 3 (1995).

[24] T.L. Mach, and J.D. Wolken, "Financial Services Used by Small Businesses: Evidence from the 2003 Survey of Small Business Finances," *Federal Reserve Bulletin* Oct.: A167-A195 (2006).

[25] The survey question specifically asked respondents about having a credit line, loan, or capital lease.

[26] K.S. Cavalluzzo and L.C. Cavalluzzo, "Market Structure and Discrimination: The Case of Small Businesses," *Journal of Money, Credit and Banking*, vol. 30, no. 4 (1998).

[27] K.S. Cavalluzzo, L.C. Cavalluzzo, and J.D. Wolken, "Competition, Small Business Financing, and Discrimination: Evidence from a New Survey," *Journal of Business*, vol. 75, no. 4 (2002).

[28] D.G. Blanchflower, P.B. Levine, and D.J. Zimmerman, "Discrimination in the Small-Business Credit Market," *The Review of Economics and Statistics*, vol. 85, no. 4 (2003).

[29] SBA officials explained that the agency defines start-up businesses as businesses in operation for less than 2 years. To make the data on conventional loans from the SSBF comparable to the SBA data, we defined a business with a conventional loan as a start-up if the business had been in operation for less than 2 years when the firm applied for the most recently approved loan.

[30] The confidence interval for the estimate of the share of conventional loans that went to small businesses in economically distressed neighborhoods (10 percent) is 7.9 to 11.7 percent.

[31] The Empowerment Zone/Enterprise Community program (EZ/EC) and Renewal Community program (RC) target federal grant monies to public and private entities, tax benefits to businesses, or both in order to improve conditions in competitively selected, economically distressed communities. For an area to be eligible for these programs at least 20 percent of the population in the census tracts that make up the area must have incomes below the national poverty line.

[32] A Historically Underutilized Business Zone is an area located in one or more qualified census tracts, qualified nonmetropolitan counties, or lands within the external boundaries of an Indian reservation.

[33] The maximum number of employees a business can have and still be considered small varies from industry to industry, but the most common standard is 500 employees. The confidence interval for the estimate of the share of conventional loans that went to small businesses with up to 5 employees (42 percent) is 38.0 to 45.2 percent, for businesses with 5 to 9 employees (24 percent) is 21.2 to 27.5 percent, and for businesses with 10 to 19 employees (17 percent) 14.0 to 19.7 percent.

[34] The confidence interval for the estimate of the share of conventional loans that went to small businesses organized as corporations (60 percent) is 56.2 to 63.5 percent, and those organized as sole proprietorships (32 percent) is 28.2 to 35.3 percent.

[35] The portfolio management score used by SBA is the Small Business Predictive Score (SBPS). The SBPS is based on consumer and business data, and assigns small businesses with scores in the absolute range of 1 to 300, but the practical range of 50 to 250. A lower score generally indicates a greater likelihood of repayment risk, while a higher score indicates a greater likelihood that the loan will be repaid.

[36] SBA says it first received SBPS credit scores for the outstanding 7(a) loans in its portfolio in March 2003. Since then, SBA has received an initial score, known as the Surrogate Origination Score, for a 7(a) loan 1 to 4 months after the loan is disbursed. SBA subsequently has received SBPS scores on a quarterly basis for almost all of the active loans in its portfolio. We obtained data for all 7(a) loans approved and disbursed from 2001 through 2005, so the dates of the initial credit scores ranged from 2003 to 2006.

[37] The earlier period of credit scores for firms that obtained credit in the conventional lending market represents data D and B/FIC had readily available and could provide us. Appendix I contains details on the data used to perform this analysis.

[38] The FSS predicts the likelihood that a business will cease operations without paying creditors in full or go into receivership.

[39] An estimated 3 percent of conventional loans had dollar amounts greater than $2 million.

[40] We used SBA data to calculate the calendar year and quarter in which each loan was approved and to calculate interest rates for all loans in a given quarter that were for under $1 million.

[41] We used the Federal Reserve's *Survey of Terms of Business Lending*, which provides information quarterly on commercial and industrial loans of loans in four size categories (less than $100,000; from $100,000 through $999,999; from $1 million through $999,999,000; and $10 million or more) made only by commercial banks. We used only data related to the first two categories because those loan amounts most resembled the 7(a) loans in the SBA data and, as discussed previously, SBA considers loans reported in call report data of $1 million or less to be for small businesses.

[42] We used the Federal Reserve's historical reports on the monthly bank prime rate to estimate the prime rate for every quarter from 2001 through 2004.

[43] GAO, Small Business Administration: Model for 7(a) Program Subsidy Had Reasonable Equations, but Inadequate Documentation

Hampered External Reviews, GAO-04-9 (Washington, D.C.. Mar. 31, 2004).

APPENDIX I

[1] For example, we eliminated records where a loan maturity date preceded or equaled the disbursement date or records in which the SBA-guaranteed percentage exceeded the maximum level allowed by the program.

[2] According to *Financial Services Used by Small Businesses: Evidence From the 2003 Survey of Small Business Finances*, about 1 percent of small businesses indicated that the government was the supplier of their financial services. Federal Reserve staff noted that this percentage may understate the incidence of 7(a) loans because, among other reasons, some respondents may have been unaware that they received an SBA-guaranteed loan.

[3] The SSBF initially selected 37,600 firms from D and B's Dun's Market Identifier file, of which 9,687 passed to the main questionnaire stage, and 4,268 firms completed their interviews, resulting in a weighted overall response rate of 32.4 percent. These firms represent 6.3 million small businesses. Firms eligible for the SSBF include for-profit, nonagricultural, nondepository institutions, nongovernment businesses in operation in December 2003 and during the interview, that also had less than 500 employees.

[4] The Bureau of Census organizes the 50 states and District of Columbia into nine regions, as follows: (1) East North Central (Ohio, Indiana, Illinois, Michigan, and Wisconsin); (2) East South Central (Kentucky, Tennessee, Alabama, and Mississippi); (3) Middle Atlantic (New York, New Jersey, and Pennsylvania) (4) Mountain (Montana, Idaho, Wyoming, Colorado, New Mexico, Arizona, Utah, and Nevada); (5) New England (Maine, New Hampshire, Vermont, Massachusetts, Rhode Island, and Connecticut); (6) Pacific (Washington, Oregon, California, Alaska, and Hawaii); (7) South Atlantic (Delaware, Maryland, District of Columbia, Virginia, West Virginia, North Carolina, South Carolina, Georgia, and Florida); (8) West North Central (Minnesota, Iowa, Missouri, North Dakota, South Dakota, Nebraska, and Kansas); and (9) West South Central (Arkansas, Louisiana, Oklahoma, and Texas).

[5] The survey does not include information on loans under $1,000.

[6] Gross loan extensions made during the first full business week in the middle month of each quarter by a sample of 348 commercial banks of all sizes. The sample data are used to estimate the terms of loans extended during that week at all insured commercial banks. The survey notes that the estimated terms of bank lending are not intended for use in measuring the terms of loans extended over the entire quarter or residing in the portfolios of those banks.

INDEX

A

B

S

T

U

V